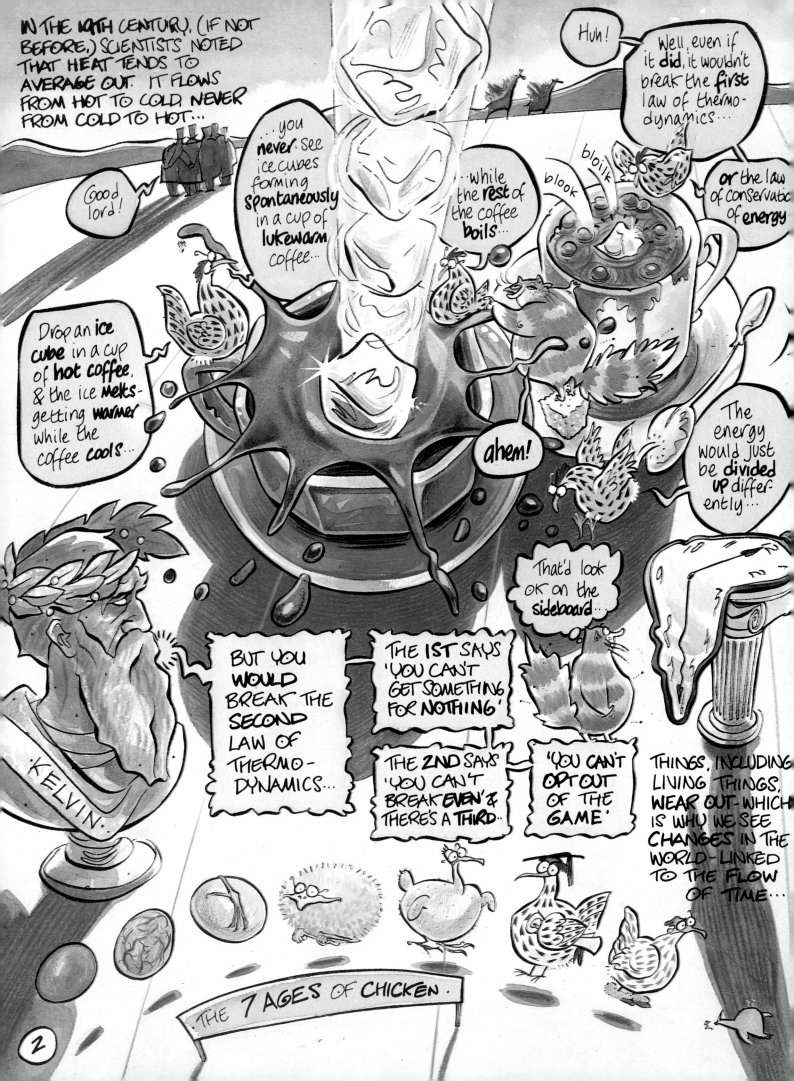

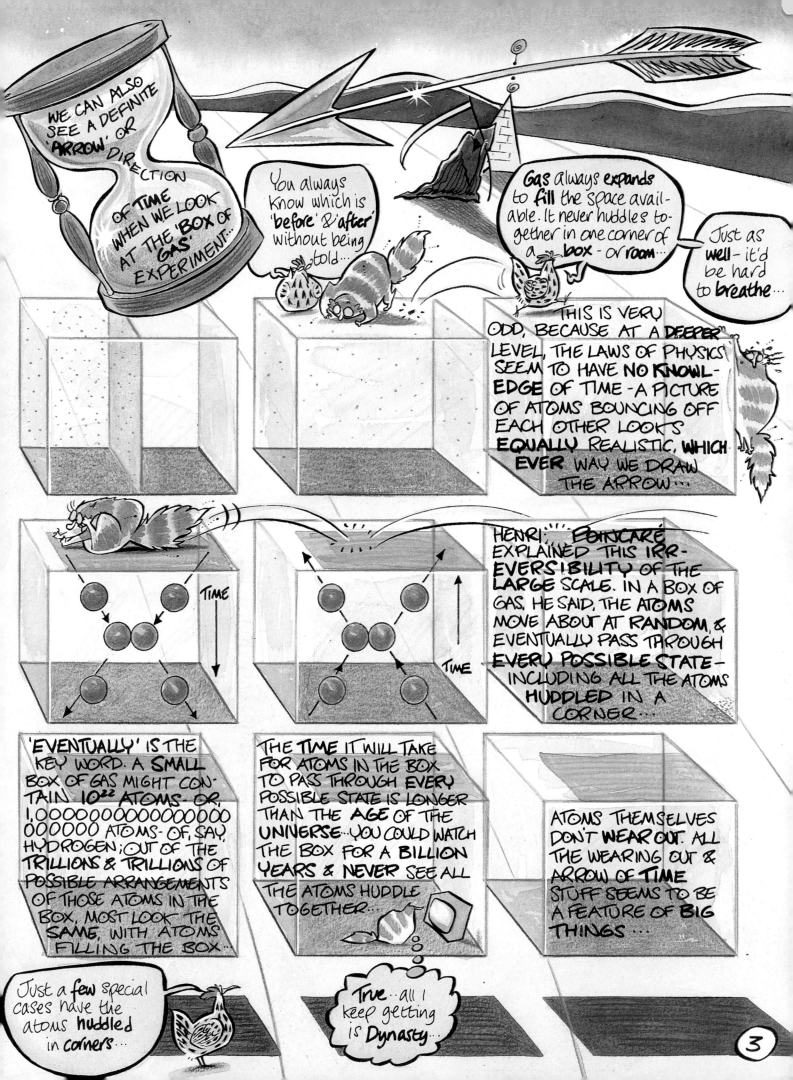

NOBEL PRIZEWINNER ILYA PRIGOGINE HAS A NEW THEORY OF TIME. HE SAYS THAT **MICROSCOPIC** SYSTEMS ARE JUST AS REVERSIBLE AS **MACROSCOPIC** SYSTEMS...

...WITH AN **INBUILT** ARROW OF TIME... HE BASES HIS IDEA ON A **COMBINATION** OF THERMODYNAMICS & QUANTUM THEORY...

could I forget?

Wait for it!

REMEMBER THE **BOX** OF GAS WITH ATOMS FLYING ABOUT & BOUNCING **OFF** EACH OTHER? OR...

WHRRRR

NEWTON'S LAWS SAY THAT EACH OF THESE COLLISIONS STILL OBEY THE LAWS OF PHYSICS IF IT IS **REVERSED** - WHICH MEANS THE WHOLE BOX OF GAS IS '**TIME REVERSIBLE**' WHICH MEANS THAT IF, BY MAGIC, THE MOTION OF **EVERY ATOM** IN THE GAS BOX COULD BE REVERSED...

....IT'S BEHAVIOUR WOULD **STILL** SATISFY THE **NEWTONIAN LAWS**...

SINCE THE 19TH CENTURY, PHYSICISTS HAVE PUZZLED HOW A WORLD BUILT OUT OF TIME REVERSIBLE **ATOMS** CAN BE IRREVERSIBLE ON THE **LARGE** SCALE. BUT **PRIGOGINE** SAYS THESE ATOMS ARE **NOT** TIME REVERSIBLE.

Think about how we **glibly** say '**reverse** the motion of **every atom** simultaneously...'

we do?

To understand, we need to know **exactly** where every atom is...

..& exactly **where**, & how **fast**, it's **going** - it's position & momentum...

RRRRRIPP

But **Quantum physics** says this is **impossible!**

THE EQUATIONS OF QUANTUM PHYSICS ARE THEMSELVES REVERSIBLE - THEY HAVE **NO** ARROW OF TIME BUILT INTO THEM, ANY MORE THAN NEWTON'S **EQUATIONS** HAVE. BUT QUANTUM PHYSICS HAVE A **NEW FEATURE** - DISCOVERED BY **WERNER HEISENBERG.**

HEISENBERG PROVED THAT IT IS IMPOSSIBLE TO **KNOW**, AT SUBATOMIC LEVELS...

POCK!

EXACTLY WHERE A PARTICLE **IS**, & WHERE IT'S **GOING**...

THIS IS KNOWN AS HEISENBERG'S **UNCERTAINTY PRINCIPLE**...

PLUNK!

6

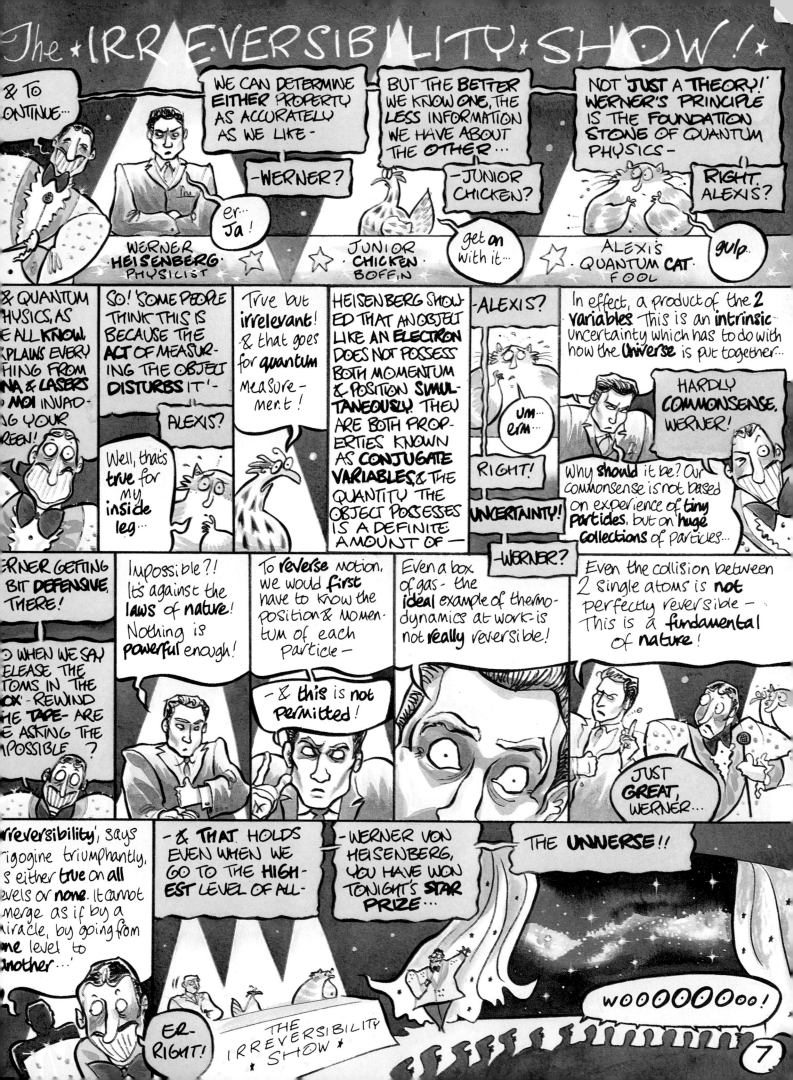

The *IRREVERSIBILITY* SHOW!*

& TO CONTINUE...

WE CAN DETERMINE **EITHER** PROPERTY AS **ACCURATELY** AS WE LIKE –
– **WERNER**?

er... Ja!

BUT THE **BETTER** WE KNOW ONE, THE **LESS** INFORMATION WE HAVE ABOUT THE **OTHER**...
– JUNIOR CHICKEN?

get on with it...

NOT '**JUST A THEORY**!' **WERNER'S** PRINCIPLE IS THE FOUNDATION STONE OF QUANTUM PHYSICS –
RIGHT, **ALEXIS**?

gulp..

WERNER **HEISENBERG** PHYSICIST

JUNIOR **CHICKEN** BOFFIN

ALEXIS QUANTUM **CAT** FOOL

& QUANTUM PHYSICS, AS WE ALL **KNOW**, EXPLAINS EVERYTHING FROM DNA & LASERS TO MOI INVADING YOUR SCREEN!

SO! 'SOME PEOPLE THINK THIS IS BECAUSE THE **ACT OF MEASURING** THE OBJECT **DISTURBS** IT' –
ALEXIS?

Well, that's **true** for MY inside leg...

True but *irrelevant*! & that goes for *quantum* measurement!

HEISENBERG SHOWED THAT AN OBJECT LIKE AN **ELECTRON** DOES NOT POSSESS BOTH MOMENTUM & POSITION **SIMULTANEOUSLY**. THEY ARE BOTH PROPERTIES KNOWN AS **CONJUGATE VARIABLES**, & THE QUANTITY THE OBJECT POSSESSES IS A DEFINITE AMOUNT OF –

– ALEXIS?

UM... erm..

RIGHT!

UNCERTAINTY!

In effect, a product of the **2 variables** This is an **intrinsic** uncertainty which has to do with how the **Universe** is put together...

HARDLY **COMMONSENSE**, WERNER!

Why **should** it be? Our commonsense is not based on experience of **tiny** particles, but on '**huge** collections of particles...

WERNER GETTING A BIT **DEFENSIVE** THERE!

SO WHEN WE SAY 'RELEASE THE ATOMS IN THE BOX' – REWIND THE TAPE – ARE WE ASKING THE IMPOSSIBLE...?

Impossible?! It's against the **laws of nature**! Nothing is **powerful** enough!

To **reverse** motion, we would **first** have to know the position & momentum of each particle –
– & **this** is not **permitted**!

Even a box of gas – the **ideal** example of thermodynamics at work – is not **really** reversible!

– **WERNER**?

Even the collision between 2 single atoms is **not** perfectly reversible – this is a **fundamental** of nature!

JUST **GREAT**, WERNER...

irreversibility,' says Prigogine triumphantly, 'is either **true** on all levels or **none**. It cannot emerge as if by a miracle, by going from one level to another...'

ER – RIGHT!

– & **THAT** HOLDS EVEN WHEN WE GO TO THE **HIGHEST** LEVEL OF ALL –

– WERNER VON HEISENBERG, YOU HAVE WON TONIGHT'S **STAR PRIZE**...

THE **UNIVERSE**!!

THE IRREVERSIBILITY SHOW

WOOOOOOOOo!

THE WAVELENGTH OF THE LIGHT HAS INCREASED ON ITS JOURNEY TO US — BECAUSE SPACETIME HAS EXPANDED WHILE IT WAS EN ROUTE.

GALAXIES DO NOT MOVE 'THROUGH' SPACE, ANY MORE THAN THE PAINT SPOTS MOVE 'THROUGH' THE FABRIC OF THE BALLOON…

THE GREATEST TRIUMPH OF GENERAL RELATIVITY IS THAT THE EQUATIONS PREDICTED THE DISCOVERY OF THE EXPANDING UNIVERSE…

…EVEN THOUGH EINSTEIN HIMSELF DID NOT AT FIRST BELIEVE THE EQUATIONS HE HAD INVENTED, & TRIED TO SWEEP THE PREDICTIONS UNDER THE CARPET…

HE LATER CALLED THIS 'THE BIGGEST MISTAKE OF MY LIFE'…

The Universe is expanding…

But in both cases, the space between them gets bigger…

The BIG BANG was the beginning of TIME!

We look out into the Universe…

…& see that the galaxies are moving apart as spacetime expands…

We do?

Do they?

12

TO GIVE EINSTEIN HIS DUE, HIS FAILURE TO BELIEVE HIS OWN EQUATIONS WASN'T REALLY **SUCH** A BIG BLUNDER- IN 1915, NOBODY KNEW THAT **OTHER** GALAXIES EXISTED...

THEY THOUGHT THAT THE UNIVERSE CONSISTED ONLY OF THE COLLECTION OF **STARS** THAT WE NOW CALL THE **MILKY WAY GALAXY**...

..& **OUR** GALAXY ITSELF IS CERTAINLY **NOT** EXPANDING, BECAUSE ON SUCH A **SMALL** SCALE AS AN INDIVIDUAL GALAXY THE **GRAVITY** OF ALL THE STARS HOLDS THE CLUSTER TOGETHER.

Aww- Stop at **Milky Way** Services, Dad...

WHAT DO WE MEAN BY **SMALL?** THE MILKY WAY GALAXY IN WHICH THE SUN IS AN **AVERAGE** SORT OF STAR, CONTAINS ABOUT **100 BILLION** STARS...

..MOSTLY SPREAD OVER A DISC SOME **100000** LIGHT YEARS IN **DIAMETER** & **2000** LIGHT YEARS **THICK**, WITH OTHER STARS FILLING A SPHERICAL **HALO 500000** LIGHT YEARS AROUND THE DISC..

(A **LIGHT YEAR** IS A MEASURE OF **DISTANCE**, NOT TIME - IT'S THE DISTANCE LIGHT TRAVELS IN ONE YEAR. LIGHT TRAVELS AT **300,000 KM** A **SECOND**. A LIGHT YEAR IS A **LONG WAY** - NEARLY TEN THOUSAND BILLION KM)

THE SUN ORBITS IN THE PLANE OF THIS DISC, ABOUT **30000** LIGHT YEARS OUT FROM THE CENTRE. (3/5 OF THE WAY TO THE EDGE) THE WHOLE KIT & CABOODLE IS HELD TOGETHER BY GRAVITY

YOU ARE HERE

Forget it, Marvin - you've wasting **no more money** on -

- bloody **Earth Colonize** machines...

IT WAS ONLY IN THE 1920s THAT AMERICAN ASTRONOMER EDWIN **HUBBLE**, USING THE NEW 100-INCH TELESCOPE AT **MOUNT WILSON** IN CALIFORNIA

(THE BIGGEST TELESCOPE IN THE WORLD UNTIL AFTER WORLD WAR II) FIRST PROVED THAT FAINT PATCHES OF LIGHT SEEN ON THE SKY ARE REALLY **OTHER** GALAXIES...

- IN MANY CASES, EVERY BIT AS **BIG** AS OUR OWN MILKY WAY - & THEN DISCOVERED THE **REDSHIFT** EFFECT...

THAT PUT GENERAL RELATIVITY ON A SECURE FOOTING AS THE BEST MATHEMATICAL DESCRIPTION OF THE EXPANDING UNIVERSE, GOT RID OF EINSTEIN'S **FIDDLE FACTOR**, & LAID THE FOUNDATIONS FOR MODERN COSMOLOGY...

I was bo-o-o-rn under a Wandering Star...

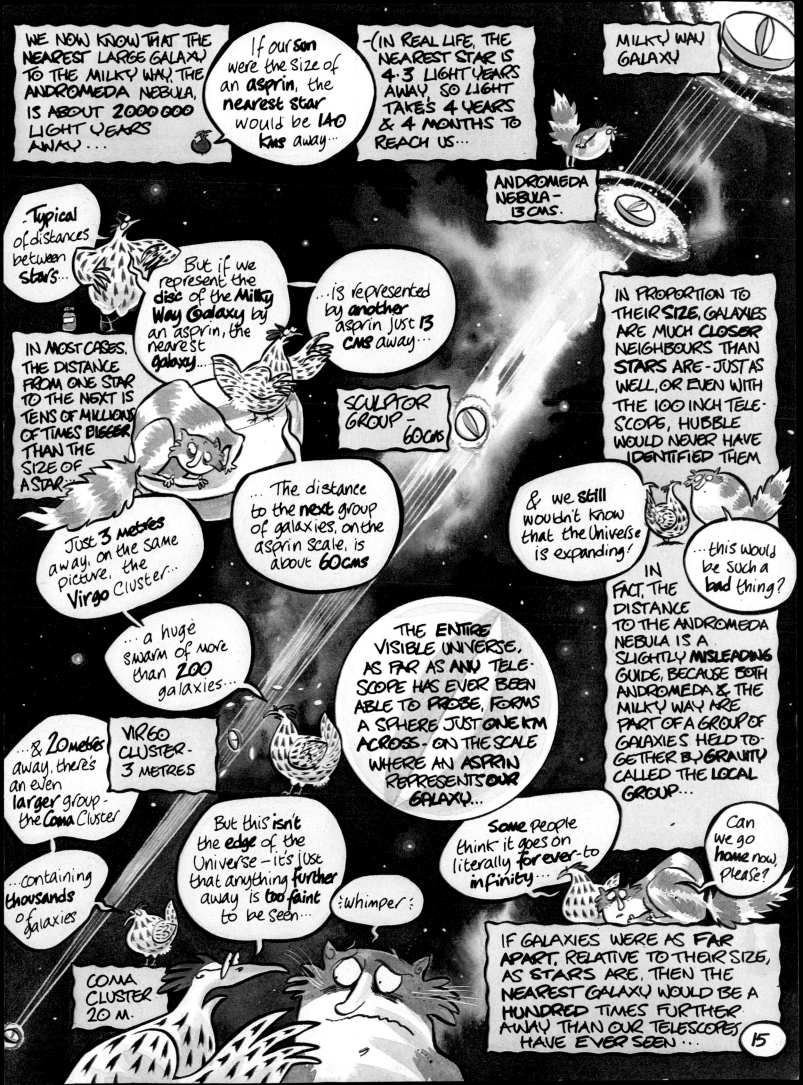

MODERN COSMO-LOGY BEGAN WITH GENERAL RELATIVITY, ALONG WITH THE PREDICTION THAT THE UNIVERSE IS EXPANDING. THE KEY IN-SIGHT FROM EINSTEIN'S WORK IS THAT SPACE ITSELF (OR SPACETIME) IS BENT...

THIS MEANS THAT THE GEOMETRY OF SPACETIME IS NOT THE SAME AS THE EUCLIDEAN GEOMETRY WE LEARNED AT SCHOOL —

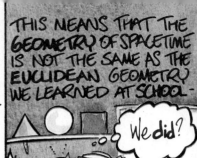

We did?

S'true!

— WHICH ONLY APPLIES ACCURATELY TO **FLAT** SURFACES

CRASH!

AS WELL AS THE DIS-TORTIONS IN SPACE-TIME PRODUCED BY LUMPS OF MATTER LIKE THE SUN...

'matter' - not 'batter'

THE OVERALL BACKGROUND SPACE OF THE UNIVERSE IS NOT ONLY EXPANDING, BUT CURVED...

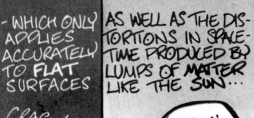

SINCE SPACE OCCUPIES 3 DIMENSIONS, THE BENDING MUST INVOLVE A 4TH DIMENSION - OR A 5TH IF YOU RECKON TIME IS THE 4TH...

MOST PEOPLE HAVE **TROUBLE** PICTURING 5 DIMENSIONS

BAM!

BUT YOU CAN GET A FEEL FOR WHAT'S GOING ON BY LOOK-ING AT WHAT HAPPENS TO GEOMETRY WHEN WE BEND A FLAT, 2-DIMENSIONAL SHEET OF PAPER...

THERE ARE 3 KINDS OF GEOMETRY THAT COULD APPLY TO SPACE. IF IT IS 'POSIT-IVELY CURVED' OR CLOSED LIKE THE SURFACE OF A SPHERE, THEN 'PARALLEL' LINES EVENTUALLY CROSS EACH OTHER - & THE ANGLES OF A TRIANGLE ADD UP TO MORE THAN 180°

CLOSED SPACE

IF IT IS 'NEGATIVELY CURVED' OR OPEN LIKE A MOUNTAIN PASS OR A **SADDLE** SURFACE, THEN PARALLEL LINES **DIVERGE** FROM ONE ANOTHER & THE ANGLES ADD UP TO **LESS THAN** 180°

OPEN SPACE

THE 3RD POSSIBILITY, FLAT SPACE, IS A SPECIAL CASE DIV-IDING THE OTHER TWO POSSIBILITIES.

IN **FLAT** SPACE, & **ONLY** IN FLAT SPACE, PARALLEL LINES ALWAYS STAY THE **SAME** DISTANCE APART, & THE ANGLES OF A TRI-ANGLE ADD UP TO **180°**, AS **EUCLID** EXPLAINED

FLAT SPACE

Tell that to Euclid...

IN **PRINCIPLE**, YOU COULD FIND OUT THAT THE EARTH IS SPHER-ICAL, & WORK OUT ITS SIZE BY DRAWING **HUGE** TRIANGLES ON THE SURFACE OF THE SAHARA DESERT & MEASURING THEIR ANGLES EXTREMELY ACCURATELY

IN 3-DIMENSIONAL SPACE, THE EQUIVALENT MEASURE-MENTS INVOLVE COUNTING THE **NUMBERS OF GALAXIES** AT DIFFERENT DISTANCES FROM US...

... measuring how the volume of the Universe increases as we look further away

THESE MEASUREMENTS ARE VERY **DIFFICULT**, BUT THEY TELL US THAT OUR UNIVERSE IS **NEARLY** FLAT. IT **MIGHT** JUST BE **OPEN** - IT **MIGHT** JUST BE **CLOSED**

oh...

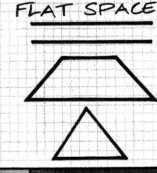

— & MIGHT BE A PROFOUND INSIGHT INTO HOW THE UNI-VERSE CAME IN-TO BEING IN THE FIRST PLACE...

WE EXPLAIN WHY IN SECTION 5... pages 42/3

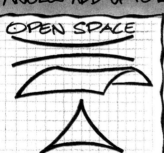

AS SOON AS **NEWS** OF GENERAL RELATIVITY ARRIVED, **SOME** CLEVER SODS SCIENTISTS USED IT TO PRODUCE THEIR **OWN** MATHEMATICAL DESCRIPTIONS OF THE UNIVERSE - OR RATHER, THE **UNIVERSES**...

- IT TURNED OUT THAT EINSTEIN'S EQUATIONS PROVIDE A DESCRIPTION OF ALL KINDS OF POSSIBLE UNIVERSES, SOME OPEN SOME CLOSED, SOME FLAT...

Some as bent as £2 notes! Get yer lovely universes!

- all quality stuff...

HONEST BOFFIN

THERE IS JUST ONE BASIC SET OF EQUATIONS - BUT WHEN YOU'VE SOLVED THEM, YOU CAN GET LOTS OF DIFFERENT ANSWERS.

Nice, eh?

...& all of 'em right!

ONE SET OF SOLUTIONS PROBABLY DESCRIBES OUR UNIVERSE - BUT WE DON'T KNOW EXACTLY WHICH ONE...

C'mon - you might PICK the lucky one!

-Make your reputation in theoretical physics!

I thought they were supposed to know...

Don't worry-they know enough to make it all worthwhile

A DUTCHMAN, WILLEM DE SITTER SOON FOUND, LIKE EINSTEIN, THAT EVEN WHEN YOU TRIED TO FIND SOLUTIONS THAT DESCRIBE A STATIC UNIVERSE, THEY COME WITH BUILT-IN EXPANSION...

Lovely, ain't they? Take a set home for the wife...

THE FIRST PERSON TO APPRECIATE THAT THE EXPANSION IS AN **INTEGRAL** PART OF THE EQUATIONS, & BELONGS IN COSMOLOGY FROM THE WORD GO - WAS A RUSSIAN, ALEXANDER FRIEDMAN...

- That's 'GO' spelt 'B-A-N-G'!

H'mm..

WHO WAS ALSO THE **FIRST** MATHEMATICIAN TO REALISE THAT HE WAS DEALING WITH A **FAMILY** OF SOLUTIONS TO EINSTEIN'S EQUATIONS...

?

FRIEDMAN SHOWED THAT IN **SOME CASES**, THE UNIVERSE EXPANDS - FOREVER...

BRAVO!

BUT IN OTHER CASES IT EXPANDS OUT TO A CERTAIN SIZE - THEN **COLLAPSES** BACK UPON ITSELF. (THE FIRST IS OPEN, THE SECOND CLOSED)

BUT THE PERSON (OR PARSON) WHO DID **MOST** AT THE END OF THE 1920s TO PUBLICISE THE IDEA THAT THE UNIVERSE HAD STARTED OUT IN A **SUPER-DENSE** STATE & HAS BEEN **EXPANDING** EVER SINCE (EVEN IF IT MIGHT ONE DAY **RECOLLAPSE**)-

-WAS A BELGIAN PRIEST GEORGE LEMAÎTRE

That's a very old joke...

THE 'FATHER OF COSMOLOGY'

HE PUT THE MATHEMATICS OF EXPANDING SPACE & THE OBSERVATIONS OF RECEDING GALAXIES **TOGETHER**, & CONCLUDED THAT LONG AGO, EVERYTHING IN THE UNIVERSE HAD BEEN **PILED UP** IN WHAT HE CALLED THE **PRIMEVAL ATOM**.

The Father of Cosmology, I presume?

HIS IMAGINATION **FAILED** HIM BEFORE HE GOT BACK TO THE **SINGULARITY** AT THE **MOMENT** OF **CREATION**

JEHOVAH'S NURSERIES
· F1 HYBRIDS
CONTS: 1 JUMBO
UNIVERSAL COSMIC
SEED

YOU WILL GET

- & HE ENVISAGED A TIME WHEN THE ENTIRE CONTENT OF THE UNIVERSE WAS PACKED INTO A SPHERE ABOUT **30 TIMES BIGGER** THAN OUR SUN...

needs no compost!

...WHICH THEN EXPLODED OUTWARDS, BREAKING UP INTO **FRAGMENTS** THAT BECAME THE ATOMS, STARS & GALAXIES WE KNOW...

MODERN COSMOLOGY STILL HAS THIS PICTURE AT IT'S **HEART** - THOUGH NOW COSMOLOGISTS CAN IMAGINE BACK TO A TIME WHEN THE UNIVERSE WAS **EVEN MORE** COMPACT & DENSE...

THE BIG BANG CALCULATIONS GAVE GAMOW, ALPHER & ANOTHER STUDENT, ROBERT **HERMAN**

Pity he wasn't called **Delter**...

- A WAY TO TAKE THE **TEMPERATURE** OF THE UNIVERSE AT THE TIME THE NUCLEAR REACTIONS THAT MADE **HELIUM** WERE GOING ON.

Not **that** *way*...

THEY REALISED THAT THE FIREBALL OF RADIATION FROM THE BIG BANG WOULD STILL **FILL** THE UNIVERSE TODAY...

...not **that** *sort of radiation*

EVEN THOUGH IT'S BEEN EXPANDING & COOLING FOR BILLIONS OF **YEARS**, THEY CALCULATED THAT IT WOULD NOW HAVE A TEMPERATURE OF ABOUT **5K** —

...MINUS DEGREES...

BUT THEY NEVER IMAGINED THAT RADIO ASTRONOMY TECHNIQUES WOULD SOON BE SENSITIVE ENOUGH TO MEASURE THE WEAK BACKGROUND HISS OF RADIATION...

I thought that was just **background noise** *on the Archers...*

That too...

THE MEASUREMENTS WERE ACTUALLY MADE IN THE **1960**s, GIVING THE TEMPERATURE OF THE RADIATION AS JUST UNDER **3K** —

-270°C

MINUS DEGREES...

- PROVIDING THE CLINCHING EVIDENCE THAT PERSUADED MOST ASTRONOMERS THAT THE BIG BANG **REALLY DID** HAPPEN...

Whimper...

- THAT THE UNIVERSE WAS INDEED BORN IN A SUPERDENSE FIREBALL ABOUT **15 BILLION YEARS** AGO...

THE RESULT WAS AN EXPLOSION OF **INTEREST** IN THE **BIG BANG** IDEA, LEADING TO NEW THEORIES ABOUT THE **BIRTH** & **DEATH** OF THE UNIVERSE...

What?

rubbish!

BUT WHEREAS THE 'OLD' BIG BANG THEORY WAS BASED PURELY ON GENERAL RELATIVITY —

Is that **copyright**?

I need to see my **agent**...

- THE NEW IDEAS TAKE ON BOARD ANOTHER GREAT DEVELOPMENT IN 20TH CENTURY SCIENCE - **QUANTUM THEORY**!

SCREAM!

What about the other **1%** *of elements, then?*

- you, me, bats, newts, **Trump Tower**...

In fact, **Fred Hoyle** *in the 1950s showed how* **all** *the* **other** *elements are actually made in the* **stars**...

INSIDE A **STAR** MIGHT NOT BE AS **HOT** AS THE BIG BANG ITSELF WAS - BUT STARS ARE AROUND FOR BILLIONS OF YEARS...

Oh, it's **hot** *all right*...

So there's **plenty** *of time for atomic nuclei to* **bash** *into one another &* **stick together**...

THE **HEAVY ELEMENTS** THEN GET SPREAD OUT THROUGH SPACE WHEN SOME STARS **EXPLODE** AS **SUPERNOVAS**...

Wow! I'll bet **he** *got the Nobel* **Prize** *for* **that**!

Actually, one of his colleagues, **Willy Fowler**, *got it*...

Like most of the scientific **establishment**, *the* **Nobel Prize** *committee had been* **annoyed** *by some of Hoyle's later & more* **cranky** *ideas*...

- like **Archeopteryx** *being a* **fake**...

H'm he wouldn't hold with **talking chickens**, *then*...

CATS & QUANTA 3

QUANTUM PHYSICS IS ABOUT CHANGE & UNCERTAINTY

Hey, look! - A Science Fair!

NEWTON'S LAW OF PHYSICS HAD SUGGESTED THAT THE UNIVERSE RAN LIKE CLOCKWORK, WOUND UP BY GOD IN THE BEGINNING...

I thought he went in for apples...

...& THAT IF YOU KNEW WHERE EVERY PARTICLE OF MATTER WAS AT ANY INSTANT, & WHICH DIRECTION IT WAS MOVING IN...

...YOU COULD PREDICT THE FUTURE FROM THE LAWS OF PHYSICS WITH ABSOLUTE PRECISION...

MADAM ZAZA NEWTONIAN PALMIST.

THIS WORRIED PHILOSOPHERS, & SEEMED TO LEAVE NO ROOM FOR FREE WILL. QUANTUM PHYSICS RETURNED UNCERTAINTY & FREE WILL TO SCIENCE...

&-yes-I see funding...

She could miss...

Or not throw at all... THUNK

I thought he wore the blindfold...

- WHICH MADE THE PHILOSOPHER HAPPY, BUT UPSET A LOT OF PHYSICISTS...

Yeah!

Look- it's fixed, you know...

THE STRANGENESS OF THE QUANTUM WORLD- QUANTUM PHYSICS REALLY ONLY APPLIES TO THINGS ABOUT THE SIZE OF, OR SMALLER THAN, ATOMS & MOLECULES...

Ha! That one's plain glass!

CAN BEST BE SEEN BY CONSIDERING THE SAGA OF THE ELECTRON...

WHAT THE PHYSICIST SAW!

IN THE LATE 19th CENTURY PHYSICISTS WERE PUZZLED BY THE NATURE OF RADIATION THAT IS EMITTED FROM A WIRE THAT CARRIES AN ELECTRIC CURRENT THROUGH A TUBE PUMPED EMPTY OF AIR.

Wow!

WERE THESE CATHODE RAYS, AS THEY WERE CALLED, A FORM OF RADIATION LIKE LIGHT OR RADIO WAVES?

'H'MM...'

OR WERE THEY IN FACT PRODUCED BY A STREAM OF TINY PARTICLES? JJ THOMSON (ALWAYS KNOWN BY HIS INITIALS, NEVER BY NAME)

'JJ'

...ALLEGEDLY THE CLUMSIEST PHYSICIST THAT EVER LIVED (WHO HAD HIS EXPERIMENTS BUILT & OPERATED BY OTHER PEOPLE)

'OUCH!'

- WORKING AT THE CAVENDISH LABORATORY IN CAMBRIDGE, PROVED IN THE 1890s THAT CATHODE RAYS ARE ACTUALLY A STREAM OF ELECTRICALLY CHARGED PARTICLES- NOW KNOWN AS ELECTRONS...

'AH-HA!'

IN 1906, HE RECEIVED THE NOBEL PRIZE IN PHYSICS FOR PROVING THAT CATHODE RAYS- ELECTRONS -ARE 'PARTICLES'...

'THANKS!'

RICHARD FEYNMAN ONCE SAID THAT THE CENTRAL MYSTERY OF QUANTUM PHYSICS IS CONTAINED IN THE 'EXPERIMENT WITH TWO HOLES'. THE CLEAREST PRACTICAL DEMONSTRATION OF WHAT HE MEANT CAME IN 1989, WHEN JAPANESE RESEARCHERS SUCCEEDED IN BUILDING AN EXPERIMENT SO SUBTLE... ...IT COULD SHOOT INDIVIDUAL ELECTRONS THROUGH A SCREEN WITH 2 HOLES IN IT & DETECT THEIR ARRIVAL ON THE OTHER SIDE ON A TV SCREEN.

H'm

IF WAVES PASS THROUGH TWO HOLES IN THIS WAY, THE WAVE PATTERN FROM ONE HOLE INTERFERES WITH THE WAVE PATTERN FROM THE OTHER HOLE— SO THAT ON THE OTHER SIDE OF THE HOLES THERE IS A SET OF INTERFERENCE FRINGES WITH THE WAVES ADDING TOGETHER IN SOME PLACES & CANCELLING EACH OTHER OUT IN BETWEEN.

SCREEN
SLIT A
SLIT B

Well!

IF A SERIES OF PARTICLES IS FIRED THROUGH 2 HOLES (FEYNMAN USED TO IMAGINE A MACHINE GUN BLASTING AWAY)— THEN THERE SHOULD BE A LOT OF PARTICLES ARRIVING ON THE OTHER SIDE OF EACH HOLE— BUT NO INTERFERENCE... THE JAPANESE FIRED ELECTRONS ONE AT A TIME FROM AN ELECTRON MICROSCOPE 'GUN' THROUGH A 'DOUBLE SLIT' —CALLED AN ELECTRON BIPRISM. WHEN EACH ELECTRON ARRIVED AT THE TV SCREEN, IT MADE A TINY SPOT OF LIGHT, DEMONSTRATING THAT IT WAS A SINGLE PARTICLE.

Logical...

Um-

-oh..

BUT AFTER THOUSANDS OF ELECTRONS HAD BEEN FIRED AT THE SCREEN ONE AT A TIME... THE PICTURE THAT HAD BUILT UP WAS ONE OF ALTERNATING BRIGHT & DARK STRIPES— AN INTERFERENCE PATTERN. EACH ELECTRON HAD BEHAVED LIKE A WAVE GOING THROUGH THE APPARATUS, PASSING THROUGH BOTH SIDES OF THE ELECTRON BIPRISM AT ONCE, & INTERFERING WITH ITSELF IN SOME WAY TO MAKE THE PATTERN.

!

Yow!

I need help!

Go on! Hit me! I'll feel better!

Can I quote you?

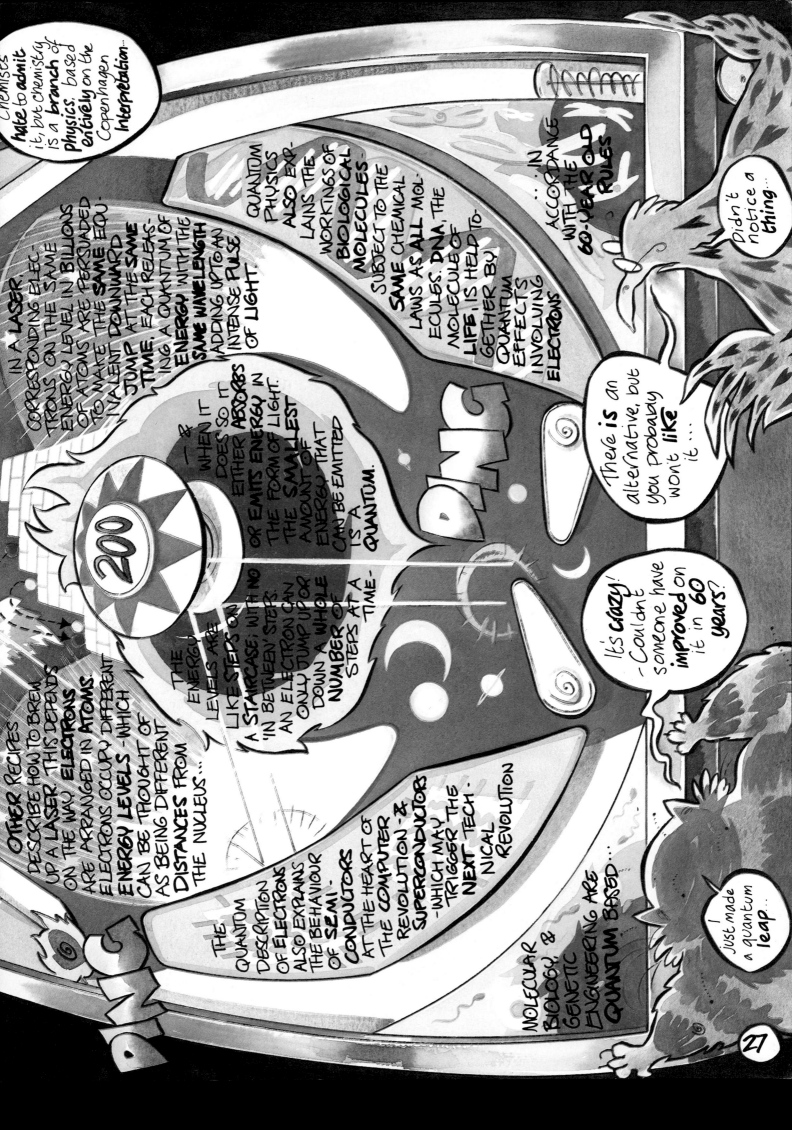

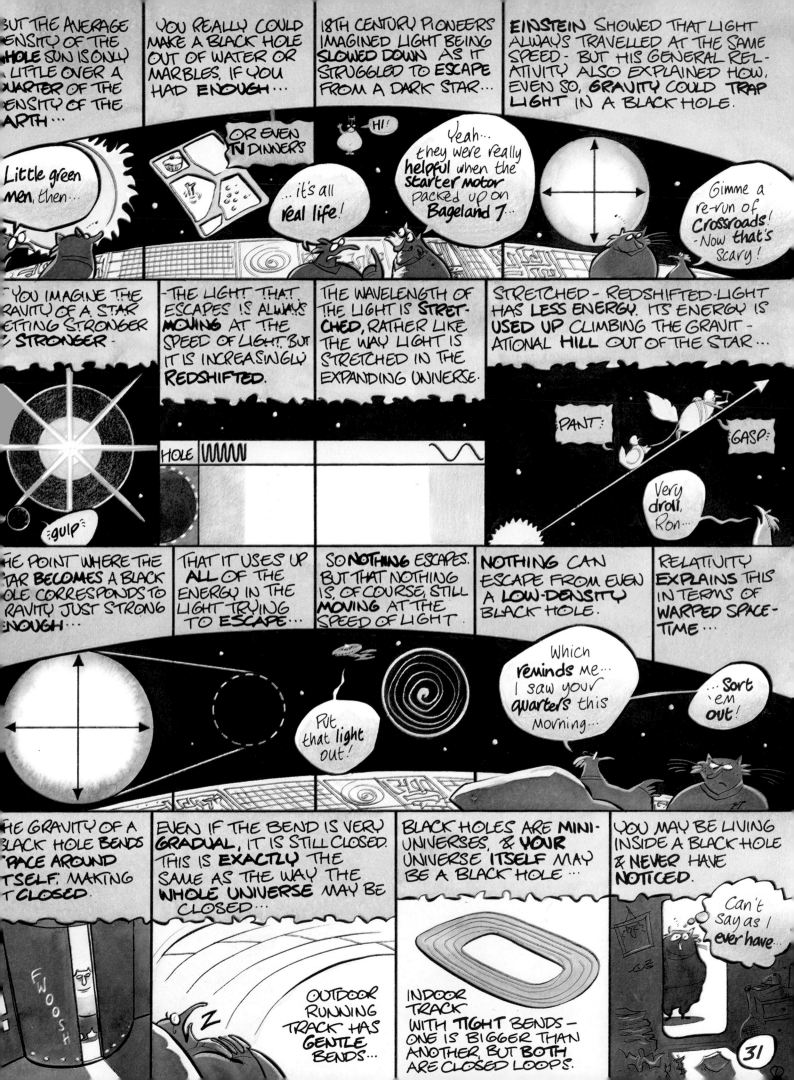

What about the **other** kind of black holes...

...the kind you see in **movies**?

THESE WERE INVENTED - OR DISCOVERED MATHEMATICALLY - BY AN **INDIAN STUDENT** IN THE EARLY 1930s...

BUT NOBODY TOOK HIS WORK **SERIOUSLY** FOR MORE THAN **30 YEARS**...

Well...it's interesting...

You reactivated it **deliberately**, didn't you?!

THE 18TH-CENTURY PIONEERS IMAGINED INCREASING THE **ESCAPE VELOCITY** FROM A STAR BY ADDING **MORE MASS**, BUT KEEPING ITS DENSITY CONSTANT...

ANOTHER WAY TO INCREASE THE ESCAPE VELOCITY IS TO KEEP THE **SAME MASS** BUT LET THE STAR **SHRINK** - SO THAT ITS DENSITY INCREASES...

Listen to it while you **muck out** the alien then!

You wanted a **pet**...

aw...

AN ORDINARY STAR LIKE YOUR SUN **DOESN'T** SHRINK, EVEN THOUGH GRAVITY IS PULLING IT TOGETHER...

I'm going to have a **good read**...

BECAUSE NUCLEAR REACTIONS KEEP IT **HOT** INSIDE...

F.S.S.T.

-& THE HEAT CREATES A **PRESSURE** WORKING AGAINST THE **INWARD TUG** OF GRAVITY...

What **happens** to a **star** at the **end** of its life?

WELL-

SUBRAHAMANYAN **CHANDRASEKHAR** - THAT INDIAN STUDENT - THOUGHT ABOUT THE PROBLEM ON THE LONG **VOYAGE** TO ENGLAND...

...WHERE HE WAS TO WORK WITH THE GREAT ASTRONOMER, ARTHUR **EDDINGTON**

Get on with it!

A PLANET LIKE THE **EARTH** IS HELD UP AGAINST THE INWARD PULL OF GRAVITY BY THE **FORCES** BETWEEN ATOMS & NUCLEI...

THE ALIEN

CHANDRASEKHAR CALCULATED THAT IF THE **SUN** ITSELF COULD BE HELD UP BY THESE QUANTUM FORCES...

Yes? Can I **help** you?

I've come to - erm...

EVEN IF IT HAD **NO** SOURCE OF NUCLEAR ENERGY - IT WOULD **SHRINK** TO BECOME A STAR NO BIGGER THAN THE EARTH...

...tidy up...

There's really **no need**...

...A COOLING CINDER KNOWN AS A **WHITE DWARF**

You could change my **library** book if you like...

BUT A STAR WITH ONLY A **LITTLE** MORE MASS THAN OUR SUN - **LESS** THAN **TWICE** AS MUCH MASS -

...Michael Moorcock's not really **me**...

- WOULD BE **SO HEAVY** THAT ITS OWN WEIGHT WOULD OVERWHELM THE QUANTUM FORCES IN ITS **HEART**...

Look - you wouldn't fancy **escaping**, would you?

Y'know... **hiding** in **air ducts** & stuff...

- THERE WOULD BE **NOTHING** TO STOP IT SHRINKING FOR EVER, INTO A POINT OF INFINITE DENSITY - A **SINGULARITY**...

then I could **chase** you, &...

ahem...

EDDINGTON WAS **UNIMPRESSED**, & SNEERINGLY COMMENTED THAT SUCH A STAR MUST PRESUMABLY **CONTRACT** UNTIL 'GRAVITY BECOMES STRONG ENOUGH TO **HOLD** THE RADIATION **IN**, & THE STAR CAN AT LAST FIND **PEACE**'...

I'm sorry - it's just...**so boring** here...

HE MEANT THE COMMENT SARCASTICALLY - BUT NOW, THAT'S EXACTLY WHAT ASTRONOMERS THINK HAPPENS TO STARS - LEFT WITH MORE THAN ABOUT TWICE AS MUCH MASS AS OUR SUN AT THE END OF THEIR LIVES...

THE STAR SHRINKS & GRAVITY GETS STRONGER, WARPING SPACE AROUND ITSELF EVER MORE TIGHTLY UNTIL EVENTUALLY IT CUTS ITSELF OFF FROM THE REST OF THE UNIVERSE ENTIRELY.

I hear what you're saying...

...but unless you're into Soto Zen...

IT HAS BECOME A SELF-CONTAINED BLACK HOLE - A CLOSED REGION OF SPACETIME...

...I can't really help you...

rats!

FOR A PARTICULAR MASS, THERE'S A CERTAIN RADIUS AROUND THE CENTRAL SINGULARITY WHICH IS CUT OFF FROM THE OUTSIDE...

...& are we going to have the black hole lecture every time the chicken forgets the date?

OM...

(IN THE CASE OF YOUR SUN, IT WOULD HAVE TO BE SQUEEZED INTO A BALL WITH A RADIUS OF JUST 3KM TO BECOME A BLACK HOLE;

Huh! & I was going to call it Rover.

om

THE EARTH WOULD HAVE TO BE SQUEEZED EVEN SMALLER WITHIN A RADIUS OF ONE CENTIMETRE)

VIRGOAN DOLLAR   EARTH

NOTHING CAN ESCAPE FROM WITHIN THIS 'EVENT HORIZON'

Better cancel the space program...

NOTHING CAN GET OUT OF A BLACK HOLE, BUT IT STILL EXERTS A GRAVITATIONAL PULL ON THE WORLD OUTSIDE - & IT WILL SWALLOW UP ANYTHING THAT PASSES TOO CLOSE...

Cosmic quicksand...

SLURP

AS IT DOES SO, ITS MASS INCREASES & ITS EVENT HORIZON GROWS OUTWARD...

BUT THIS KIND OF STELLAR MASS BLACK HOLE IS A MESSY EATER...

Pardon...

MATERIAL FALLING INTO THE HOLE IS RIPPED APART BY INTENSE TIDAL FORCES & FORMS A SWIRLING DISC OF MATTER SLOWLY FUNNELING ACROSS THE EVENT HORIZON...

IN THE DISC, GRAVITATIONAL ENERGY IS CONVERTED INTO HEAT AS ATOMS BASH INTO ONE ANOTHER -

LIKE ON THE LA FREEWAY, REALLY

SO ALTHOUGH THE BLACK HOLE ITSELF IS INVISIBLE, THE REGION JUST OUTSIDE IT MAY BE RADIATING ENERGY

Hi there, rock freaks! Let's get all 3 heads banging to... Betelguse Thrash! Yeah!

- VIOLENTLY ACROSS THE ELECTRO-MAGNETIC SPECTRUM, FROM X RAYS & GAMMA RAYS TO RADIO WAVES...

HEY!

IN THE 1960s & 1970s ASTRONAUTS IDENTIFIED MANY INTENSE SOURCES OF ENERGY IN THE UNIVERSE, PINPOINTS ON THE SKY THAT BLAST OUT X RAYS INTO SPACE...

THE ONLY WAY TO EXPLAIN THESE WAS TO REVIVE CHANDRASEKHAR'S 50-YEAR-OLD THEORY

MANY OF THESE SOURCES OF ENERGY CAN ONLY BE SITES WHERE A BLACK HOLE OF THE KIND EDDINGTON DISMISSED AS NONSENSE

ORBITS AROUND AN ORDINARY STAR, STRIPPING GAS FROM IT THROUGH TIDAL FORCES -

- & SWALLOWING IT IN A VIOLENT DISPLAY OF STELLAR CANNIBALISM...

I need some air! - Or space, anyway...

...AS WELL AS THESE **ORDINARY** BLACK HOLES, THERE ARE ALSO **MINI** & **MAXI** VARIETIES...

ALMOST EXACTLY LIKE THE ONES MITCHELL ENVISAGED MORE THAN **200** YEARS AGO.

MANY GALAXIES SHINE VERY BRIGHTLY IN THEIR CENTRAL **CORES**, RADIATING NOT JUST **VISIBLE** LIGHT, BUT **RADIO WAVES**, & SHOWING OTHER SIGNS OF DISTURBANCE...

UM... *Actually, Ron,* I was hoping for a *quiet, 'emptiness of space' type* breath of air...

SORRY, LT. ALEXIS... PROGRAMS ARE THERE TO BE **PLAYED**...

CS-1

IN SOME CASES, THE BRIGHT **NUCLEUS** OF A GALAXY FAR OUTSHINES THE LIGHT OF **ALL** THE HUNDRED BILLION STARS THAT SURROUND IT **PUT TOGETHER**...

SUCH BRIGHT GALACTIC NUCLEI CAN BE SEEN ACROSS **VAST** DISTANCES OF SPACE...

SO THAT THE LIGHT YOU SEE THEM BY IS ENORMOUSLY **REDSHIFTED** IN THE EXPANDING UNIVERSE.

THE STARS OF THE GALAXY ARE TOO **FAINT** TO BE SEEN SO FAR AWAY.

-BUT THE CENTRAL **CORE** SHINES SO BRIGHTLY THAT WHEN IT IS PHOTOGRAPHED BY TELESCOPES ON EARTH, IT LOOKS LIKE A **SINGLE STAR**.

SUCH OBJECTS ARE KNOWN AS QUASISTELLAR OBJECTS OR **QUASARS**.

Now look, Ron...

BONK

CS-1

THE REDSHIFT OF A QUASAR REVEALS ITS **DISTANCE**, BECAUSE THE REDSHIFT IS **BIGGER** THE FURTHER LIGHT TRAVELS ACROSS **EXPANDING SPACE**...

THE **DISTANCE** TELLS ASTRONOMERS HOW BRIGHT A QUASAR MUST BE IN ORDER TO BE VISIBLE ON **EARTH**.

IN ORDER TO SHINE SO BRIGHT, THE POWERHOUSE OF A QUASAR IS **SWALLOWING UP** ONE OR TWO TIMES THE MASS OF YOUR SUN EVERY **YEAR**

-EATING WHOLE STARS NOT JUST STRIPPING BITS **OFF** THEM...

& CONVERTING MATTER INTO ENERGY IN LINE WITH EINSTEIN'S EQUATION $E=mc^2$. THE **GRAVITY** OF A BLACK HOLE PROVIDES THE **MOST** EFFICIENT WAY TO CONVERT **MASS** TO **ENERGY** LIKE THIS.

Hey - as you're **out there** - check out the **phasar** tubes...

& bring the milk in

34

THE MASS YOU **NEED** FOR THE **BLACK HOLE** TO **EXPLAIN** THE **POWER** PRODUCED BY A TYPICAL **QUASAR** IS ABOUT 100,000,000 TIMES THE MASS OF YOUR SUN - ALMOST **EXACTLY** THE KIND OF BLACK HOLE **MITCHELL** IMAGINED.

& I'm **sick to death** of 'put the **cat out**' jokes...

SUCH A BLACK HOLE DOES NOT INVOLVE ANYTHING GETTING **SQUEEZED** DOWN TO THE POINT WHERE QUANTUM FORCES ARE **OVERWHELMED** BY GRAVITY.

(AT LEAST, NOT TO **START** WITH, BUT ONCE IT FORMS, MATTER WILL **INVARIABLY** SETTLE INTO THE **CENTRE**)

IT COULD ARISE JUST BECAUSE **TOO MANY** STARS GOT TOO CLOSE TOGETHER IN THE CORE OF THE GALAXY...

SORTA COSMIC OSCAR SHOW...

:AHEM: - A HUNDRED MILLION SOLAR MASSES SOUNDS A **LOT**...

- BUT COMPARE IT WITH THE **MASS** OF THE **WHOLE GALAXY** - WHICH IS AROUND A HUNDRED BILLION SOLAR MASSES

THE **CENTRAL** BLACK HOLE THAT POWERS A QUASAR REPRESENTS JUST 0.1% OF THE MASS AVAILABLE IN THE SURROUNDING GALAXY...

THE **EVENT** HORIZON OF SUCH A BLACK HOLE IS NO BIGGER ACROSS THAN YOUR SOLAR SYSTEM, EVEN THOUGH IT CONTAINS THE MASS OF 100,000,000 SUNS

...& ON THE EVENT HORIZON **ITSELF** THE DENSITY IS NO GREATER...

THAN THE DENSITY OF **WATER** THAT COMES OUT OF A **TAP** IN YOUR **KITCHEN**

- IF YOU **HAD** SUCH A THING, LIEUTENANT -

AN INTREPID ASTRONAUT COULD **PILOT** A SPACECRAFT **ACROSS** THE EVENT HORIZON WITHOUT DIFFICULTY, & STUDY WHAT **HAPPENS** INSIDE A **BLACK HOLE**.

Well, as **long** as it's not us...

It's **not**, is it?

UNFORTUNATELY, THE SPACEFARER COULD **NEVER** ESCAPE, OR EVEN SEND A MESSAGE BACK TO THE OUTSIDE WORLD TO **REPORT** THOSE DISCOVERIES TO INTERESTED ASTROPHYSICISTS...

...Is it?

& WITHIN A FEW HOURS, BOTH COSMONAUT & SPACECRAFT WOULD FIRST BE **STRETCHED OUT** BY TIDAL FORCES - LIKE TOOTHPASTE BEING SQUEEZED FROM A **TUBE**...

& THEN **CRUSHED** INTO A STATE OF **ZERO VOLUME** & **INFINITE DENSITY** IN THE **CENTRAL SINGULARITY**

IS IT??

COSY, NO?

SCREAM!

? !

I've **just** had the strangest dream...

SCREAM!

I thought this was a pretty good **outfit** for the **para-scientists ball**...

ONE OF THE PEOPLE WHO PROVED THAT BLACK HOLES MUST CONTAIN A **SINGULARITY** — A POINT WHERE OUR UNDERSTANDING OF SPACE, TIME & PHYSICS **BREAKS DOWN**...

SCIENCE-AM BREAKFAST CHANNEL

So early...

WAS STEPHEN **HAWKING** OF THE UNIVERSITY OF CAMBRIDGE

...BUT HE ALSO DID SOMETHING **EVEN MORE CLEVER**... HE **COMBINED** GENERAL RELATIVITY QUANTUM THEORY & THERMODYNAMICS IN **ONE PACKAGE**

...LINKING THE BEHAVIOUR OF **BLACK HOLES** TO THE NATURE OF TIME, & **PROVING** THAT IN **SOME** CASES AT LEAST —

Black holes **ain't** so black!

IN PRINCIPLE, **ANY** AMOUNT OF MATTER, NO MATTER HOW **SMALL**, COULD MAKE A BLACK HOLE IF YOU **SQUEEZED** IT HARD ENOUGH...

PRODUCING SUCH AN **INTENSE** GRAVITATIONAL FIELD THAT IT WARPED SPACE-TIME **AROUND** ITSELF

IN FACT, EVEN **SUPERMAN CANNOT SQUEEZE THAT** HARD —

Aw, **nuts**! Diamonds **again**!

IT'S IMPOSSIBLE TO IMAGINE SQUEEZING THE **EARTH** DOWN INTO A SPHERE ICM ACROSS — A MASS OF I KILOGRAM WOULD CORRESPOND TO A BLACK HOLE ONLY I BILLIONTH AS BIG AS AN **ATOM**

Please

THE **LESS** MASS AN OBJECT HAS, THE **HARDER** IT IS TO MAKE IT INTO A BLACK HOLE...

BUT ONE PLACE WHERE MATTER WAS **REALLY** SQUEEZED HARD WAS IN THE **BIG BANG** ITSELF.

ANYTHING COULD HAVE COME OUT OF THE SUPER-DENSITY THAT EXISTED NEAR THE **SINGULARITY** IN WHICH OUR UNIVERSE WAS **BORN**...

INCLUDING **MINI** BLACK HOLES WITH A MASS OF A FEW KILOGRAMS CONTAINED WITHIN **EVENT HORIZONS** MUCH SMALLER THAN ATOMS.

YOU MIGHT THINK THAT NOTHING WOULD BE HARDER TO DETECT — BUT HAWKING PROVED THAT **WRONG** IN THE 1970s...

GENERAL RELATIVITY COMES IN, BECAUSE **GRAVITY** IS INCLUDED...

THERMODYNAMICS COMES IN, BECAUSE HAWKING SHOWED THAT EVERY BLACK HOLE HAS A **TEMPERATURE**...

— & THAT IT **KNOWS WHICH WAY** TIME FLOWS.

Bacon yes — eggs no — I know it's illogical...

ssssss

QUANTUM PHYSICS COMES IN, BECAUSE IT TELLS YOU HOW TO MAKE **PARTICLES** OUT OF **NOTHING** AT ALL...

OK — I'll demonstrate...

MAKING MATTER OUT OF ENERGY IS **EASY** — IF THERE'S ENOUGH E AROUND, YOU GET mc^2 — THE ONLY THING NATURE HAS TO WORRY ABOUT IS THAT **EACH** PARTICLE THAT IS **MADE** IS ACCOMPANIED BY ITS **ANTIPARTICLE** — A KIND OF MIRROR IMAGE

THE FIRST PERSON TO REALISE THE IMPORTANCE OF THIS WAS THE SOVIET PHYSICIST ANDREI SAKHAROV IN THE LATE 1960s...

BUILDING ON HIS WORK, PHYSICISTS NOW THINK THEY KNOW WHAT HAPPENED.

Come back, this is important!

IT TURNS OUT THAT THE LAWS OF PHYSICS ARE NOT PERFECTLY SYMMETRICAL, BUT CONTAIN A FAVOURITISM FOR MATTER OVER ANTI-MATTER.

! *HI!*

THE FAVOURITISM IS TINY, JUST ONE PART IN A BILLION. FOR EVERY BILLION ANTI-BARYONS MANUFACTURED OUT OF ENERGY, THERE WILL BE A BILLION & ONE BARYONS...

THE IMBALANCE IS FAR TOO SMALL TO HAVE SHOWN UP IN EXPERIMENTS ON EARTH...

Who are you? Another antimarlene?

BUT IT MEANS THAT IN THE COSMIC FIREBALL OF THE BIG BANG ITSELF, THERE WERE A BILLION & ONE BARYONS FOR EVERY BILLION ANTIBARYONS...

You mean you can't tell?

WHEN THE UNIVERSE EXPANDED & COOLED, EVERY ANTIBARYON EVENTUALLY MET ITS BARYONIC PARTNER...

I'm an AntiMinelli!

...& CONVERTED INTO ENERGY (ANY PROTON WILL ANNIHILATE WITH ANY ANTIPROTON, NOT A SPECIFIC PARTNER)

Paff!

JUST ONE IN A BILLION BARYONS WERE LEFT- EACH ACCOMPANIED BY A BILLION (10^9) PHOTONS.

...Sigh...

THE FACT THAT THERE ARE A BILLION PHOTONS OF THE COSMIC BACKGROUND RADIATION FOR EVERY BARYON IN THE UNIVERSE...

What good...

DEPENDS ON THE WAY THE LAWS OF PHYSICS OPERATE UNDER CONDITIONS FAR MORE EXTREME THAN ANY THAT EXIST IN THE UNIVERSE TODAY—

is sitting...

—BACK WHEN THE DENSITY OF THE ENTIRE UNIVERSE WAS GREATER THAN THE DENSITY OF THE NUCLEUS OF AN ATOM.

alone in your room?

THE ULTIMATE TEST OF THEORIES OF PHYSICS TODAY IS NOT PROVIDED BY ATOM SMASHING EXPERIMENTS.

Come...

...hear the music play...

BUT BY COMPARING WHAT THOSE THEORIES TELL US ABOUT THE EARLY UNIVERSE WITH THINGS WE CAN OBSERVE TODAY—

—SUCH AS THE RATIO OF BARYONS TO PHOTONS.

Life is a cabaret, old chum

THE WHOLE UNIVERSE IS THE PHYSICIST'S LABORATORY.

Come to the cabaret!

If you can't beat 'em...

41

At the end of the 1970s, cosmologists were puzzled by several features of the Universe...

it's very, very smooth...

WE LOOK OUT IN ONE DIRECTION (USING RADIO TELESCOPES) & CAPTURE A PHOTON THAT HAS BEEN 15 BILLION YEARS ON ITS JOURNEY TO US

ALLOWING FOR SUCH IRREGULARITIES AS GALAXIES, THE UNIVERSE LOOKS MUCH THE SAME FROM PLACE TO PLACE, & IN ANY DIRECTION YOU LOOK...

...the best way to see this is to think in terms of the cosmic background radiation

Why?

PARTICLE PHYSICS RESOLVES BOTH PUZZLES. THE WAY THAT ENERGY IS CONVERTED INTO MATTER IN THE FIRST SPLIT SECOND OF THE BIG BANG...

EACH OF THESE CHANGES RELEASES ENERGY KNOWN AS LATENT HEAT, BECAUSE THE MOLECULES IN THE WATER ARE REARRANGED IN A DIFFERENT FORM...

(...ULTIMATELY LEADING TO AN EXCESS OF ONE BARYON FOR EVERY BILLION ANTI-BARYON PAIRS) IS A KIND OF PHASE TRANSISION.

THE BEST KNOWN PHASE TRANSITIONS ARE WHEN WATER VAPOUR TURNS INTO LIQUID (CONDENSATION) OR WATER TURNING INTO ICE (FREEZING)

IN FAR LESS THAN THE BLINK OF AN EYE, A REGION OF SPACE 10^{-17} OF THE SIZE OF AN ATOM WAS INFLATED INTO A REGION THE SIZE OF A GRAPEFRUIT...

IN 10^{-33}s IT DOUBLED 10 TIMES; IN 10^{-32}s IT INCREASED IN SIZE 2^{100} TIMES, BY A FACTOR OF ABOUT 10^{50}— (A ONE FOLLOWED BY 50 ZEROS)

THEN ALL THE LATENT HEAT HAD BEEN RELEASED & THE PROCESS STOPPED...

...EVER SINCE, THE EXPANSION OF THE UNIVERSE HAS BEEN SLOWING DOWN...

IN THE SAME WAY, COSMIC INFLATION STRETCHED & SMOOTHED OUT ANY WRINKLES IN SPACETIME.

NO MATTER HOW CURVED SPACE WAS TO START WITH, AFTER STRETCHING IT BY A FACTOR OF 10^{50} IT'S VERY FLAT INDEED...

THINK OF A TINY CHILD'S BALLOON; OBVIOUSLY, ITS SKIN IS CURVED - NOW INFLATE THAT BALLOON UNTIL IT IS MANY TIMES BIGGER THAN THE EARTH

YOU CAN WALK ABOUT ON ITS SURFACE - & NEVER NOTICE ANY CURVATURE

OK- but just keep your claws in!

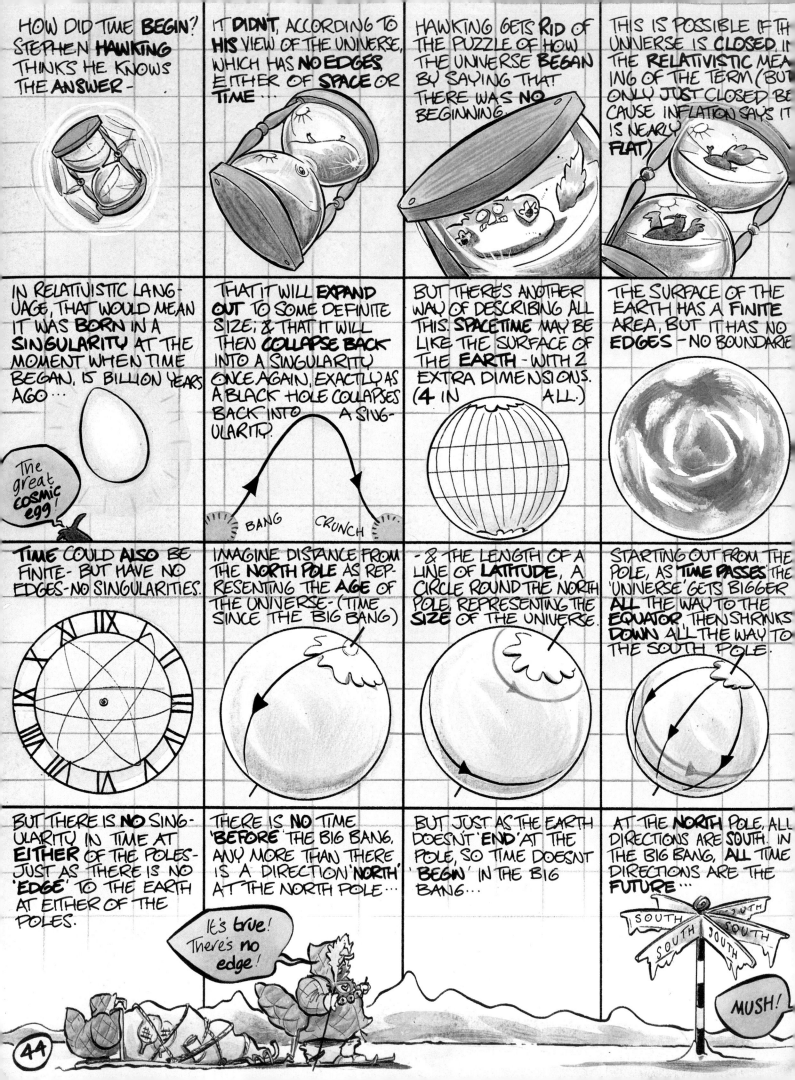

HOW DID TIME **BEGIN**? STEPHEN **HAWKING** THINKS HE KNOWS THE **ANSWER** -

IT **DIDN'T**, ACCORDING TO **HIS** VIEW OF THE UNIVERSE, WHICH HAS **NO EDGES**, EITHER OF **SPACE** OR **TIME** ...

HAWKING GETS **RID OF** THE PUZZLE OF HOW THE UNIVERSE **BEGAN** BY SAYING THAT THERE WAS **NO** BEGINNING.

THIS IS POSSIBLE IF THE UNIVERSE IS **CLOSED**, IN THE **RELATIVISTIC** MEANING OF THE TERM (BUT ONLY JUST CLOSED BECAUSE INFLATION SAYS IT IS NEARLY **FLAT**)

IN RELATIVISTIC LANGUAGE, THAT WOULD MEAN IT WAS **BORN IN A SINGULARITY** AT THE MOMENT WHEN TIME BEGAN, 15 BILLION YEARS AGO ...

The great COSMIC egg!

THAT IT WILL **EXPAND OUT** TO SOME DEFINITE SIZE; & THAT IT WILL THEN **COLLAPSE BACK** INTO A SINGULARITY ONCE AGAIN, EXACTLY AS A BLACK HOLE COLLAPSES BACK INTO A SINGULARITY.

BANG   CRUNCH

BUT THERE'S ANOTHER WAY OF DESCRIBING ALL THIS. **SPACETIME** MAY BE LIKE THE SURFACE OF THE **EARTH** - WITH 2 EXTRA DIMENSIONS. (4 IN ALL.)

THE SURFACE OF THE EARTH HAS A **FINITE** AREA, BUT IT HAS NO EDGES - NO BOUNDARIES

TIME COULD **ALSO** BE FINITE - BUT HAVE NO EDGES - NO SINGULARITIES.

IMAGINE DISTANCE FROM THE **NORTH POLE** AS REPRESENTING THE **AGE** OF THE UNIVERSE - (TIME SINCE THE BIG BANG)

- & THE LENGTH OF A LINE OF **LATITUDE**, A CIRCLE ROUND THE NORTH POLE, REPRESENTING THE **SIZE** OF THE UNIVERSE.

STARTING OUT FROM THE POLE, AS **TIME PASSES** THE 'UNIVERSE' GETS BIGGER **ALL THE WAY** TO THE **EQUATOR**, THEN SHRINKS **DOWN** ALL THE WAY TO THE SOUTH POLE.

BUT THERE IS **NO** SINGULARITY IN TIME AT **EITHER** OF THE POLES - JUST AS THERE IS NO '**EDGE**' TO THE EARTH AT EITHER OF THE POLES.

THERE IS **NO** TIME '**BEFORE**' THE BIG BANG, ANY MORE THAN THERE IS A DIRECTION '**NORTH**' AT THE NORTH POLE ...

It's **true**! There's **no** edge!

BUT JUST AS THE EARTH DOESN'T '**END**' AT THE POLE, SO TIME DOESN'T '**BEGIN**' IN THE BIG BANG ...

AT THE **NORTH** POLE, ALL DIRECTIONS ARE SOUTH. IN THE BIG BANG, **ALL** TIME DIRECTIONS ARE THE **FUTURE** ...

SOUTH   SOUTH   SOUTH   SOUTH

MUSH!

THEORIES THAT EXPLAIN WHAT HAPPENED **BEFORE** THE BIG BANG & HOW THE UNIVERSE CAME INTO BEING IN THE **FIRST** PLACE (INFLATION, QUANTUM COSMOLOGY - ACTUALLY PIONEERED BY JAYANT NARLIKAR & COLLEAGUES IN BOMBAY IN THE EARLY 1980s - & THE **BUBBLE** THEORY) TOGETHER TELL US THAT THE UNIVERSE IS JUST CLOSED...

...BECAUSE THE GRAVITY FROM ALL THE BRIGHT STARS IN ALL THE VISIBLE GALAXIES PUT TOGETHER...

ASTRONOMERS' CLUB

Extra! Universe just closed!

GASP!

This came as a **shock** to many astronomers...

...IS NOT **ENOUGH** TO CLOSE SPACE-TIME AROUND THE UNIVERSE & MAKE IT A BLACK HOLE.

Thank god...

IN **FACT**, THE **BRIGHT** STUFF PROVIDES ONLY ABOUT 10% OF THE **MASS** NEEDED TO **DO** THE JOB. BUT **SOME** ASTRONOMERS WERE **NOT** SURPRISED.

FOR **50** YEARS, THERE HAD BEEN EVIDENCE FROM THE WAY **GALAXIES** MOVE THAT THEY ARE BEING **TUGGED** BY **STRONGER** GRAVITATIONAL FORCES THAN CAN BE **EXPLAINED** BY ADDING UP THE EFFECTS OF BRIGHT STARS.

IF SO, THERE MUST BE **UNSEEN MATTER** IN THE UNIVERSE **DOING** THE **TUGGING**...

YANK

THE OBSERVATIONS WERE **IGNORED** PARTLY OUT OF PREJUDICE. EVEN IN THE 1970s...

Flying in the face of Nature!

MANY ASTRONOMERS SIMPLY DIDN'T WANT TO **BELIEVE** THAT THEY HAD SPENT THEIR LIVES STUDYING ONLY **10%** OF THE UNIVERSE...

Serves him right for writing 'Man-Master of the Universe'!

& THAT THE **OTHER** 90% WOULD NEVER **SHOW UP** IN THE VIEW THROUGH THEIR **TELESCOPES**.

& what percentage shows up through **that**?

THIS PREJUDICE WAS REINFORCED BY THE STANDARD MODEL OF THE BIG BANG-

See? Even the **new stuff** says so!

WHICH EXPLAINS SO **NEATLY** HOW MUCH HYDROGEN & HELIUM THERE IS IN THE UNIVERSE

It's an allegory the H & H percentage the Univer...

THE STANDARD MODEL **ALSO** SAYS THAT THE **COMBINED** MASS OF HYDROGEN & HELIUM, & THAT **ANY** ELEMENTS MADE OUT OF **H** & **He** INSIDE STARS...

-(THE **TOTAL** BARYONIC MASS OF THE UNIVERSE)- SHOULD BE ABOUT 10% OF THE AMOUNT NEEDED TO MAKE IT **CLOSED**.

THE STANDARD MODEL SAID THAT **BARYONS** ACCOUNT FOR 10% OF THE **CLOSURE MASS**, & **BRIGHT STARS** & GALAXIES ADD UP TO 10% OF THE CLOSURE MASS...

SO THERE WAS NO **ROOM** FOR EXTRA MASS, **WHATEVER** THE OBSERVATIONS SAID - UNLESS IT WAS SOME OTHER, **NON BARYONIC** FORM.

Which bit is the **gap**?

ES & NO. IT OULD HAPPEN. NT IF IT DID, E WOULDN'T VEN KNOW BOUT IT...

Promise?

REMEMBER THAT ANY NEW UNIVERSE MUST BE CLOSED IN THE RELATIVISTIC SENSE —

In other words - a black hole!

IT IS POSSIBLE TO IMAGINE CREATING THE SEED OF A NEW UNIVERSE ARTIFICIALLY...

Wait for it...

USING HYDROGEN BOMBS TO COMPRESS A REGION OF SPACE-TIME DOWN TO THE NECESSARY SIZE...

How about that for playing God?

BUT IF YOU DID, THE BLACK HOLE THAT YOU'D CREATED WOULD NOT START EXPLODING BACK IN YOUR FACE, SMASHING ITS WAY INTO OUR UNIVERSE...

Probably means they're trying it!

STEAD, IT EXPANDS IN SET OF DIRECTIONS AT IGHT ANGLES TO OUR MILIAR DIMENSIONS F SPACE & TIME.

OUR 3 DIMENSIONS ARE AT RIGHT ANGLES TO EACH OTHER. TIME CAN BE THOUGHT OF AS A 4TH DIMENSION AT RIGHT ANGLES TO ALL 3. THE NEW UNIVERSE HAS ITS OWN SET OF 4 DIMENSIONS, ALL OF THEM AT RIGHT ANGLES TO ALL THE DIMENSIONS OF OUR SPACETIME, & SO ON...

OUR UNIVERSE MIGHT HAVE BEEN DELIBERATELY CREATED IN THIS WAY, ALTHOUGH THERE IS NO WAY TO FIND OUT...

the multi-dimensional calculator...

SINCE EXACTLY THE SAME THING WILL HAPPEN TO ALL SEED UNIVERSES FORMED BY NATURAL FLUCTUATIONS.

Saved by cosmic geometry

I think...

OU CAN PICTURE THIS Y THINKING OF OUR NIVERSE AS THE KIN OF AN EXPANDING ALLOON GETTING IGGER AS THE BALLON IS PUMPED UP.

Here!

COSMIC FOAM KIT

A 'NEW' UNIVERSE CREATED WITHIN THAT SPACETIME (THE SKIN OF THE BALLOON IS LIKE A LITTLE BUBBLE, PINCHED OUT OF THE SKIN OF THE BALLOON, THAT EXPANDS IN IT'S OWN RIGHT...)

TWEAK

QUANTUM COSMOLOGY ALLOWS THE POSSIBILITY OF CREATING NOT JUST ONE UNIVERSE

BUT AN INFINITE NUMBER OF UNIVERSES OUT OF NOTHING AT ALL...

The ecologically safe way!

NTERCONNECTED N SOME COMPLEX, ULTI DIMENSIONAL AY - BUBBLES LIKE AM ON A RIVER...

...or frogspawn

EITHER ON HAWKING'S PICTURE, OR ON THIS PICTURE, THERE IS NO 'BEGINNING' OF TIME IN THE COSMIC SENSE

BUT TO HUMAN BEINGS THE BIG BANG IS A PRETTY DEFINITE BEGINNING.

Pretty definite for us chickens...

& FROM OUR HUMAN PERSPECTIVE, EVEN THOUGH THE FROTH ON THE NEVER ENDING RIVER OF TIME MAY EXTEND INDEFINITELY,

WHEN OUR OWN LITTLE BUBBLE COLLAPSES, THAT WILL BE THE END.

But how- & when?!

THRRRP

IN HAWKING'S UNIVERSE, THE ARROW OF TIME POINTS OUTWARD FROM THE 'NORTH POLE'

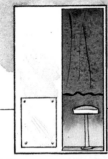

THE FUTURE IS THE DIRECTION IN WHICH SUCCESSIVE LINES OF LATITUDE GET BIGGER.

THIS REPRESENTS THE GROWTH OF ENTROPY AS THE UNIVERSE EXPANDS AWAY FROM THE NEAT SIMPLICITY OF THE BIG BANG -

& BECOMES MORE DIS-ORDERED AS STARS & BLACK HOLES FORM, CRUMPLING UP LOCAL REGIONS OF SPACETIME -

WITH THE OTHER SIDE OF THE 'EQUATOR' CORRESP-ONDING TO THE TIME WHEN THE UNIVERSE REACHES ITS MAXIMUM SIZE.

IF WE KEEP MOVING IN THE SAME DIRECTION, AWAY FROM THE NORTH POLE,

SUCCESSIVE LINES OF LATITUDE GET SMALLER, ALL THE WAY DOWN TO THE SOUTH POLE ...

THERE, ALL THE MESS & CONFUSION IN THE UNIVERSE IS WIPED AWAY...

AS EVERYTHING IS SWALLOWED UP INTO A 'NEW', SMOOTH, SINGULARITY - THE MIRROR IMAGE OF THE BIG BANG...

THIS SINGULARITY AT THE END OF TIME IS SOMETIMES REFERRED TO AS THE BIG CRUNCH - OR MORE SOBERLY, AS THE OMEGA POINT -

IT LOOKS AS IF THE UNIVERSE IS SHRINK-ING & ENTROPY IS DECREASING IN THE OTHER HALF OF SPACE-TIME...

IN OTHER WORDS, TIME RUNS BACKWARDS...

LIKE NEWTON'S LAWS OF MOTION, THE EQUATIONS THAT DESCRIBE THE BEHAVIOUR OF ELECTROMAGNETIC RADIATION HAVE **NO** INBUILT ARROW OF TIME.

D'you think I'm a bit over-dressed?

THEY WORK JUST AS **WELL** 'BACKWARDS' AS THEY DO 'FORWARDS' BUT THIS **DOESN'T** FIT OUR **COMMONSENSE** VIEW OF REALITY, & HOW WAVES BEHAVE.

IMAGINE A **STONE** DROPPED IN A **STILL** POND...

SPLOOT

RIPPLES SPREAD OUT FROM THE POINT OF IMPACT & EVENTUALLY DIE AWAY.

Sassenach vandals...

WE NEVER SEE RIPPLES BUILD UP FROM THE **EDGE** OF THE POND—

MOVE INWARDS & **COMBINE** TO FLING A STONE HIGH IN THE AIR...

YET THE ELECTROMAGNETIC EQUATIONS (DEVELOPED BY THE **SCOT** JAMES **CLERK MAXWELL**, IN THE 19TH CENTURY) SAY THE EQUIVALENT **OUGHT** TO HAPPEN...

Aye...

—TO THE KIND OF WAVES BROADCAST BY **TV** & **RADIO** STATIONS & THE **LIGHT** EMITTED BY STARS

Probing into the **Mysteries** of the Universe, eh?

A **TV MAST** CARRIES A VARYING ELECTRIC CURRENT & RADIATES ELECTROMAGNETIC WAVES IN ALL DIRECTIONS

Field trip?

IT TAKES **TIME** FOR THE WAVES, TRAVELLING AT THE **SPEED OF LIGHT**, TO REACH SOME POINT **AWAY** FROM THE MAST, SO THEY ARE CALLED THE 'RETARDED' WAVES

Confused? Frustrated? All Too Much?

MAXWELL'S EQUATIONS DESCRIBE PERFECTLY THE WAY WAVES **PROPAGATE**. BUT THEY DO **MORE**...

& so do we!

...The James Clerk Maxwell **Wave-Particle Sheep!**

LIKE DIRAC'S DISCOVERY OF A **SECOND** SET OF SOLUTIONS TO THE EQUATIONS THAT DESCRIBE **MATTER**, MAXWELL'S EQUATIONS HAVE A **SECOND** SET OF SOLUTIONS.

—wanna see 'em?

THESE DESCRIBE WAVES **CONVERGING** FROM **SPACE** ONTO THE TV MAST, & PRODUCING A **VARYING** ELECTRIC CURRENT IN IT.

SUCH WAVES CROSS SPACETIME **FAR AWAY** FIRST—THEY ARE CALLED 'ADVANCED' WAVES.

No bad, eh?

WE NEVER SEE ADVANCED WAVES IN **OUR** UNIVERSE, & THIS IS **ANOTHER** MANIFESTATION OF THE **ARROW OF TIME**.

But why?

ANOTHER WAY OF DESCRIBING ADVANCED WAVES IS TO SAY THAT THEY MOVE 'OUTWARDS' FROM THE MAST, BUT 'BACKWARDS' IN TIME.

52

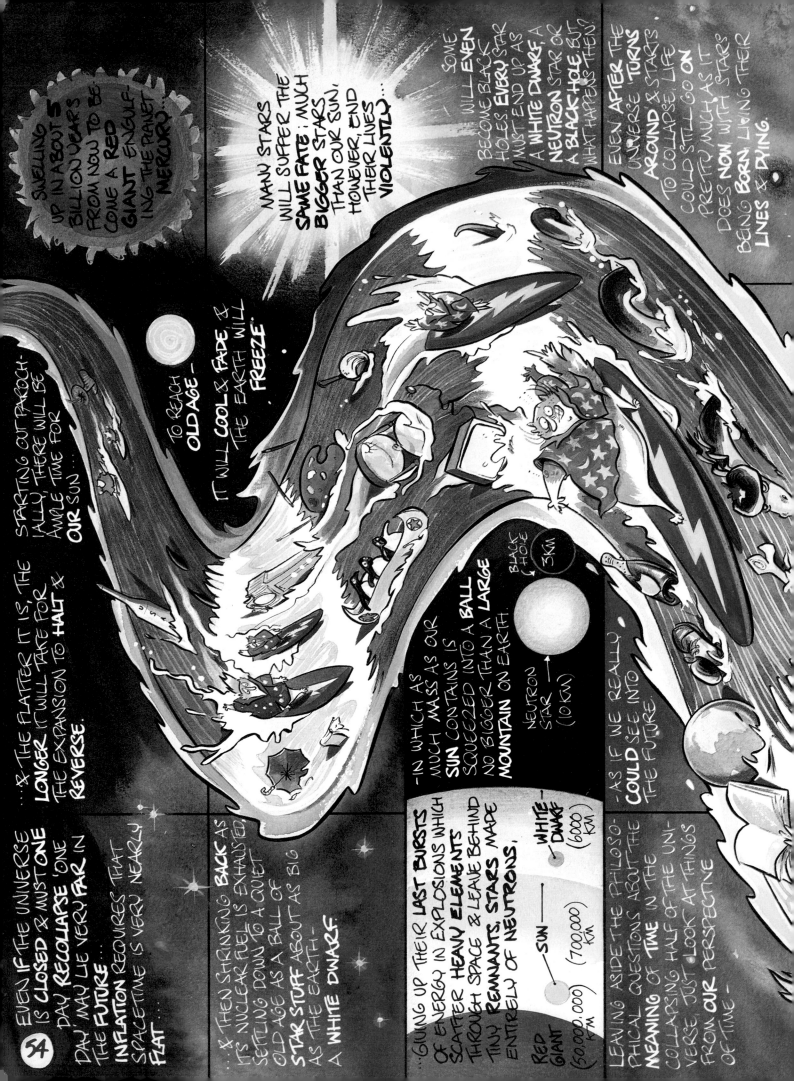

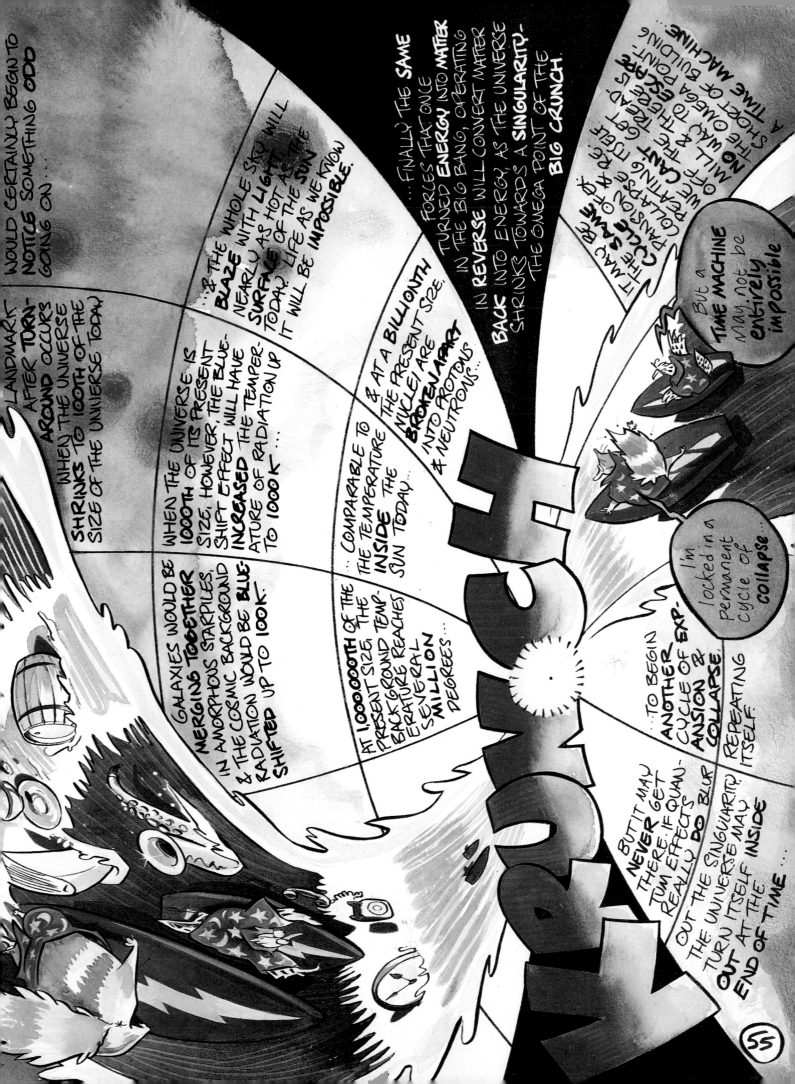

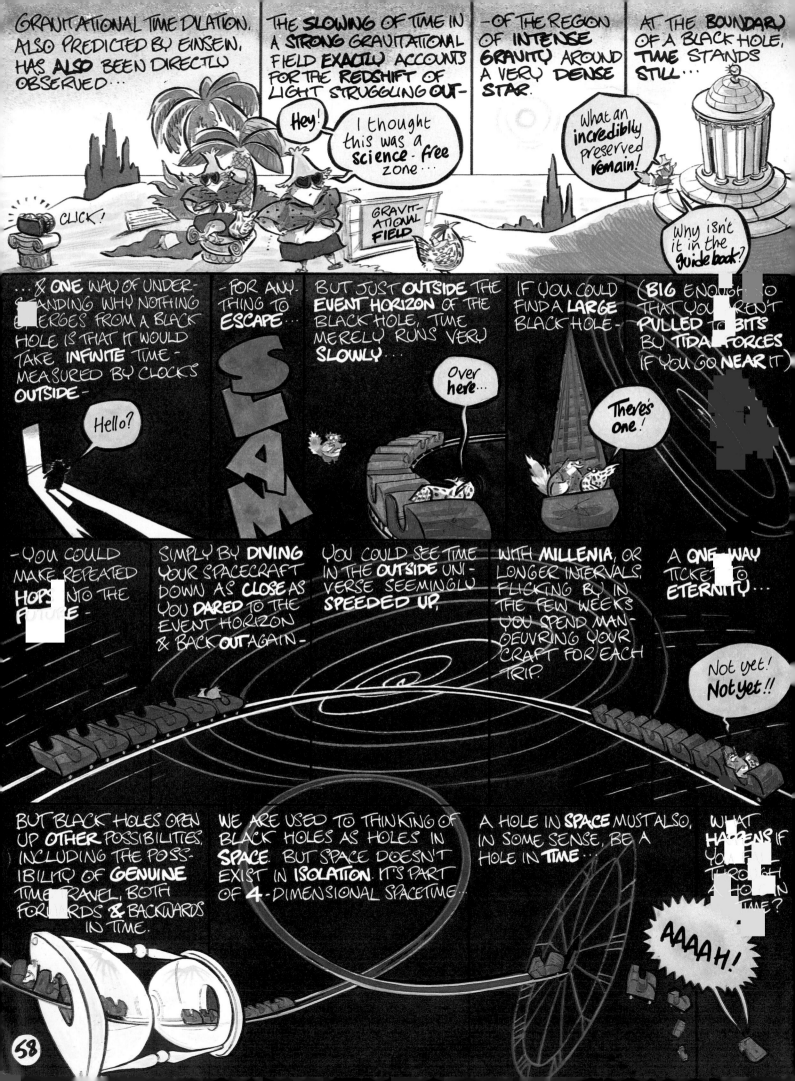

GRAVITATIONAL TIME DILATION, ALSO PREDICTED BY EINSTEIN, HAS **ALSO** BEEN DIRECTLY OBSERVED...

THE **SLOWING** OF TIME IN A **STRONG** GRAVITATIONAL FIELD **EXACTLY** ACCOUNTS FOR THE REDSHIFT OF LIGHT STRUGGLING OUT—

Hey!

I thought this was a science-free zone...

GRAVITATIONAL FIELD

CLICK!

—OF THE REGION OF **INTENSE** GRAVITY AROUND A VERY DENSE STAR.

AT THE BOUNDARY OF A BLACK HOLE, TIME STANDS STILL...

*What an **incredibly** preserved remain!*

Why isn't it in the guidebook?

... & **ONE** WAY OF UNDERSTANDING WHY NOTHING EMERGES FROM A BLACK HOLE IS THAT IT WOULD TAKE **INFINITE** TIME— MEASURED BY CLOCKS OUTSIDE—

Hello?

—FOR ANYTHING TO **ESCAPE**...

SLAM

BUT JUST **OUTSIDE** THE **EVENT HORIZON** OF THE BLACK HOLE, TIME MERELY RUNS VERY SLOWLY...

Over here...

IF YOU COULD FIND A **LARGE** BLACK HOLE—

There's one!

(**BIG** ENOUGH SO THAT YOU AREN'T PULLED TO BITS BY TIDAL FORCES IF YOU GO **NEAR** IT)

—YOU COULD MAKE REPEATED **HOPS** INTO THE FUTURE—

SIMPLY BY **DIVING** YOUR SPACECRAFT DOWN AS **CLOSE** AS YOU **DARED** TO THE EVENT HORIZON & BACK **OUT** AGAIN—

YOU COULD SEE TIME IN THE **OUTSIDE** UNIVERSE SEEMINGLY **SPEEDED UP**,

WITH **MILLENIA**, OR LONGER INTERVALS, FLICKING BY IN THE FEW WEEKS YOU SPEND MANOEUVRING YOUR CRAFT FOR EACH TRIP.

A **ONE-WAY** TICKET TO **ETERNITY**...

Not yet! Not yet!!

BUT BLACK HOLES OPEN UP **OTHER** POSSIBILITIES, INCLUDING THE POSSIBILITY OF **GENUINE** TIME TRAVEL, BOTH FORWARDS & BACKWARDS IN TIME.

WE ARE USED TO THINKING OF BLACK HOLES AS HOLES IN **SPACE**. BUT SPACE DOESN'T EXIST IN **ISOLATION**. IT'S PART OF **4**-DIMENSIONAL SPACETIME...

A HOLE IN **SPACE** MUST ALSO, IN SOME SENSE, BE A HOLE IN **TIME**...

WHAT HAPPENS IF YOU GO THROUGH A HOLE IN TIME?

AAAAH!

THE BLACK HOLES WE HAVE DISCUSSED SO FAR ARE NON-ROTATING. THEY JUST SIT THERE IN SPACE...

SWALLOWING UP ANYTHING THAT COMES TOO CLOSE & CRUSHING IT INTO THE CENTRAL SINGULARITY, BUT THIS ISN'T REALLY VERY REALISTIC.

No? Thank god...

JUST ABOUT EVERYTHING IN THE UNIVERSE ROTATES: THE EARTH, SUN, OTHER PLANETS & OTHER STARS

& LIKE A SKATER WHO SPINS FASTER WHEN SHE PULLS HER ARMS IN, ANYTHING THAT SHRINKS WILL SPIN FASTER...

EEEEEEEEEEEEEK!

THIS, FOR THOSE WHO REALLY WANT TO KNOW, IS THE LAW OF CONSERVATION OF ANGULAR MOMENTUM: BIG THINGS HAVE MORE MOMENTUM, & THIS TURNS INTO FASTER SPIN WHEN THEY SHRINK.

Big deal...

A ROTATING BLACK HOLE IS DIFFERENT FROM A NON-ROTATING BLACK HOLE IN ONE PARTICULARLY IMPORTANT WAY—

INSTEAD OF HAVING A SINGLE EVENT HORIZON, & A POINT OF SINGULARITY AT ITS CENTRE,

A ROTATING BLACK HOLE HAS TWO EVENT HORIZONS, WITH SPACETIME AROUND THE SUN REGULARLY DISTORTED IN SUCH A WAY—

THAT IT OPENS A GATEWAY TO ANOTHER REGION OF SPACETIME — IN EFFECT, THROUGH THE SINGULARITY & OUT THE OTHER SIDE...

Both my event horizons are about to radically distort...

STARTING FROM EVERYDAY SPACE, A TRAVELLER COULD, IN PRINCIPLE, PASS THROUGH BOTH EVENT HORIZONS, EMERGING INTO ANOTHER REGION OF SPACE.

BUT NOT JUST ANOTHER REGION OF SPACE...

...another region of spacetime!

SUCH A JOURNEY COULD TAKE YOU INTO THE FUTURE OR THE PAST, OR EVEN INTO ANOTHER UNIVERSE ALTOGETHER (PERHAPS THERE'S A WAY TO ESCAPE THE BIG CRUNCH!)

MOST PHYSICISTS, & OTHER PEOPLE, DRAW BACK IN HORROR AT THE PROSPECT

& I'm here to tell them they're right!

1066 2001 3510 76 1666 1962

GENUINE TIME TRAVEL RAISES ALL KINDS OF PARADOXES, ALONG THE LINES OF A TRAVELLER WHO GOES BACK IN TIME & KILLS HER OWN GRANNY BEFORE HER MOTHER IS BORN...

Where did the time traveller come from?

BUT, PARADOXES OR NO, THERE IS NOTHING IN THE LAWS OF PHYSICS THAT FORBIDS TIME TRAVEL...

FOR TRAVEL SICKNESS

& OUR COMMON SENSE HAS BEEN PROVED FAULTY AT LEAST TWICE.

IT MAY, THOUGH, BE VERY DIFFICULT TO BUILD A PRACTICAL, WORKING TIME MACHINE.

What the hell — think big!

LET'S LOOK AT HOW IT MIGHT BE DONE — GIVEN UNLIMITED RESOURCES...

In 1974, Frank Tipler of Tulane University, New Orleans, caused a flurry in scientific dovecotes...

...by publishing a paper in the highly respectable **PHYSICAL REVIEW** in which he proved that time travel **IS** possible.

*Or at least, **not forbidden** by the laws of physics...*

RUBBISH HOLIDAY READING

BEST OF PHYSICAL REVIEW

UNDEMANDING NOVEL

HEAVY STUFF

First, he showed that equations allow in theory for the existence of **JOURNEYS** through spacetime which **RETURN TO** their **STARTING POINT**...

*You **unnatural fowl!** I'm off to do **holiday** stuff!*

...in both space & time '**CLOSED TIMELIKE LOOPS**' in which part of the journey must be **BACKWARDS** in time...

You're ridiculous!

Then, he proved that the conditions required for closed timelike loops could arise **NATURALLY** in the universe.

*Manifest through a **time warp**, did they?*

Finally, he showed **HOW** it is possible, in **PRINCIPLE**, to construct an **ARTIFICIAL** time machine...

Look...

It all depends on the way massive **ROTATING OBJECTS** distort spacetime...

Dragging the dimensions of spacetime **AROUND** with the rotation, & making the dimensions '**TIP OVER**' in a 4-dimensional sense...

Three axes of space & one of time are all, remember, perpendicular to each other.

FORWARD

RIGHT

UP

A traveller who keeps moving 'forward'

In the strong gravitational field region near a very dense **ROTATING OBJECT**, however,

F

U

R

*...as they pass into a **strong** field...*

The **DRAG** effect tips these 4 perpendicular directions so that the direction that **USED** to represent time becomes a **SPACE** direction —

F

U

R

actually ends up moving 'right'...

While one of the **SPACE** dimensions becomes a **TIME** dimension.

*'Time' here being more a an **axis** in the **mind***

U

R

F

In **PRACTICAL** terms, the roles of **SPACE** & **TIME** have been **REVERSED**.

Not very romantic though, is it?

In a cerebral sort of way...

A traveller moving through space in this **DISTORTED** region of spacetime is moving **THROUGH TIME** in the world outside —

C'mon... we'll have a 'Time Machine'!

— where gravity is **WEAK** — such a traveller could journey into the distorted region,

Cheers!

Backwards in time & back out into the everyday world without **EVER** violating the rules of **RELATIVITY**.

*It's all **true!***

IN ORDER TO DO THE TRICK, THOUGH, YOU NEED A **LOT OF MATTER**, ABOUT 10 TIMES THE MASS OF THE SUN, ROTATING VERY FAST, **TWICE EVERY 1000TH OF A SECOND** - THE DENSITY HAS TO BE SO GREAT THAT A **SINGULARITY** FORMS AT THE CENTRE.

IF THE 'TIME MACHINE' WERE **NOT ROTATING** - THAT WOULD MEAN IT WAS A **BLACK HOLE** SO THE TRAVELLER COULD **NEVER** GET BACK INTO THE OUTSIDE AFTER ALL...

> Don't **stop!**

BUT VERY FAST ROTATION ALSO HAS THE EFFECT OF **FLINGING AWAY** THE EVENT HORIZON...

> Lose your inhibitions & event horizons!

> 2 more!

- LEAVING A '**NAKED**' SINGULARITY BEHIND. - ONE YOU CAN TRAVEL **CLOSE TO** & STILL **ESCAPE FROM**...

SUCH AN OBJECT **COULD** OCCUR NATURALLY IF A **NEUTRON** STAR WERE TO **SPIN FAST** ENOUGH.

> That one's doing it!

> They're **all** doing it!

'ALL' THAT YOU **NEED** TO **MAKE** A TIME MACHINE IS A **CYLINDER** AS **DENSE** AS A NEUTRON STAR, ABOUT **10 KM ACROSS** & **100 KM LONG** -

SPINNING **TWICE EVERY MILLISECOND**, SO THAT THE **RIM** OF THE CYLINDER IS MOVING AT **HALF** THE SPEED OF LIGHT.

WEAK FIELD — STRONG FIELD — WEAK FIELD

(ACTUALLY, THIS IS LIKE **10 NEUTRON STARS** JOINED END TO END - OR POLE TO POLE...)

MANUFACTURING SUCH AN OBJECT MIGHT PROVE INSURMOUNTABLY **DIFFICULT**. BUT NOT LONG AFTER TIPLER CAME UP WITH THE DESCRIPTION OF A WORKING TIME MACHINE,

> TIME MACHINE HOLDINGS, INC. ©
> - KEY RINGS!
> - BLACK HOLE COASTERS!
> - SWALLOWS YOUR DRINK PROBLEM!

ASTRONOMERS IDENTIFIED A FEW RAPIDLY ROTATING NEUTRON STARS, KNOWN AS **PULSARS**, THAT **REALLY DO** SPIN ONCE EVERY ONE OR TWO MILLISECONDS.

> GULP!

> Black Hole **Boxers?!**

NO **RESPECTABLE** SCIENTIST HAS DARED TO MAKE THE CONNECTION. BUT IT IS ASTONISHING THAT OBJECTS WHICH BEAR SUCH A **CLOSE RESEMBLANCE** TO THE RELATIVISTIC DESCRIPTION OF A WORKING TIME MACHINE **REALLY DO EXIST IN OUR UNIVERSE.**

IF WE COULD FIND ONE **NEARBY**, & ENCOURAGE IT TO SPIN JUST A LITTLE BIT **FASTER**,

> MY OTHER STAR'S A PULSAR

WE WOULD BE ABLE TO **TEST** TIPLER'S IDEAS & FIND OUT WHETHER TIME TRAVEL IS **POSSIBLE.**

> Read the book? Seen the video? Worn the T-shirt?

EVEN **THAT** IS A DAUNTING PROSPECT. THE **GOOD** NEWS IS THAT YOU ONLY HAVE TO DO THE TRICK FOR A **FLEETING** INSTANT,

> - Sick of the **present**? Dogged by **past** mistakes?

SINCE AT THE MOMENT OF ITS **CREATION** THE TIME MACHINE IS TIED **FOREVER** TO ALL CLOSED TIMELIKE LINES INTO THE **FUTURE.**

> Try a Getaway **one-way** ticket!

> TIME TRAVEL AGENCY

YOU CAN NEVER GO BACK IN TIME **EARLIER** THAN THE MOMENT THAT THE TIME MACHINE **EXISTED**...

> But even if it did only **exist** for a **moment**, it opens up the **entire future** for exploration!

WITH **UNLIMITED** ENGINEERING ABILITY, THOUGH, A SPACEFARING CIVILIZATION **COULD CONSTRUCT** AN EVEN **MORE** USEFUL KIND OF TIME MACHINE...

IN **1989**, RESEARCHERS AT CALTECH (HEADED BY KIP **THORNE**) & MOSCOW STATE UNIVERSITY (HEADED BY IGOR NOVIKOV) JOINED FORCES TO AMAZE THEIR COLLEAGUES WITH A **NEW** VARIATION ON THE THEME...

They suggest using **WORMHOLES** to travel through time!

A **WORMHOLE**, IN RELATIVE THEORY, IS RATHER LIKE A BLACK HOLE WHICH HAS NOT **QUITE** COLLAPSED INTO A SINGULARITY –

BUT OPENS OUT AGAIN INTO A **SEPARATE REGIO** OF SPACETIME.

It's rather like...

...THE **WARDROBE** IN C.S. LEWIS'S THE LION, THE WITCH & THE WARDROBE... IN **OUR** UNIVERSE, YOU GO THROUGH A SMALL **DOOR**...

BUT THE **BACK** OF THE WARDROBE OPENS OUT INTO A **DIFFERENT** UNIVERSE – NARNIA.

A WORMHOLE CAN OPEN UP INTO **ANOTHER** UNIVERSE, OR CAN JOIN TWO REGIONS OF OUR OWN UNIVERSE.

Yeah. But **that's** just in **books**!

THE **PROBLEM** IS THAT ACCORDING TO RELATIVITY THEORY **ALONE** WORMHOLES SHOULD VERY RAPIDLY **COLLAPSE** INTO SINGULARITIES...

Hey...

BEFORE ANYONE HAS A CHANCE TO TRAVEL **THROUGH** THEM...

Yow!

BUT IT IS POSSIBLE THAT **QUANTUM** EFFECTS CAN KEEP A WORMHOLE OPEN RATHER LIKE THE WAY THEY **SMEAR OUT** THE SINGULARITY AT THE BEGINNING OF TIME

IT WOULD BE A SMALL WORMHOLE – BUT NEVER MIND.

eeek!

DRINK ME

But **how** would it be used for **time travel**?

To **stop** anybody being **late**–

WHAT YOU NEED IS A WORMHOLE **JOINING** TWO BLACK HOLES NEXT TO EACH OTHER IN **OUR** UNIVERSE...

It – it's the **White Chicken**!

NOW, TAKE BLACK HOLE B & WHIRL IT ROUND WORMHOLE A AT NEARLY THE SPEED OF LIGHT...

Don't quibble – this is the **easy** bit!

Black holes have **strong** gravity, OK?

But...

SO JUST DANGLE A NICE BIG ASTEROID IN FRONT OF IT, & THE BLACK HOLE WILL COME RUNNING.

SHOOT THE ASTEROID ROUND & ROUND IN A CIRCLE –

& THE HOLE WILL CHASE IT LIKE A DOG CHASING ITS OWN TAIL...

Simple, like I said!

PLUME
Published by the Penguin Group
Penguin Books USA Inc., 375 Hudson Street, New York, New York 10014, U.S.A.
Penguin Books Ltd, 27 Wrights Lane, London W8 5TZ, England
Penguin Books Australia Ltd, Ringwood, Victoria, Australia
Penguin Books Canada Ltd, 2801 John Street, Markham, Ontario, Canada L3R 1B4
Penguin Books (N.Z.) Ltd, 182-190 Wairau Road, Auckland 10, New Zealand

Penguin Books Ltd, Registered Offices: Harmondsworth, Middlesex, England

Published by Plume, an imprint of New American Library, a division of Penguin Books USA Inc.
First published by MacDonald & Co. (Publishers) Ltd , London.

First Plume Printing, September, 1990
10 9 8 7 6 5 4 3 2 1

Copyright © Kate Charlesworth and John Gribbin, 1990
All rights reserved. For information address New American Library.

Ⓟ

REGISTERED TRADEMARK—MARCA REGISTRADA
ISBN 0-452-26495-2

Printed in the United States of America